Thermal Power Plants

Modeling, Control, and Efficiency Improvement

Thermal Power Plants

Modeling, Control, and Efficiency Improvement

Xingrang Liu
Ramesh Bansal

CRC Press
Taylor & Francis Group
Boca Raton London New York

CRC Press is an imprint of the
Taylor & Francis Group, an **informa** business

MATLAB® is a trademark of The MathWorks, Inc. and is used with permission. The MathWorks does not warrant the accuracy of the text or exercises in this book. This book's use or discussion of MATLAB® software or related products does not constitute endorsement or sponsorship by The MathWorks of a particular pedagogical approach or particular use of the MATLAB® software.

CRC Press
Taylor & Francis Group
6000 Broken Sound Parkway NW, Suite 300
Boca Raton, FL 33487-2742

First issued in paperback 2020

© 2016 by Taylor & Francis Group, LLC
CRC Press is an imprint of Taylor & Francis Group, an Informa business

No claim to original U.S. Government works

ISBN 13: 978-0-367-57470-3 (pbk)
ISBN 13: 978-1-4987-0822-7 (hbk)

Library of Congress Cataloging-in-Publication Data

Names: Liu, Xingrang. | Bansal, Ramesh C.
Title: Thermal power plants : modeling, control, and efficiency improvement / authors, Xingrang Liu and Ramesh Bansal.
Description: Boca Raton : Taylor & Francis, CRC Press, 2016. | Includes bibliographical references.
Identifiers: LCCN 2015048816 | ISBN 9781498708227 (alk. paper)
Subjects: LCSH: Electric power plants--Computer simulation. | Heat engineering. | Electric power plants--Fuel.
Classification: LCC TK1041 .L58 2016 | DDC 621.31/21--dc23
LC record available at http://lccn.loc.gov/2015048816

Visit the Taylor & Francis Web site at
http://www.taylorandfrancis.com

and the CRC Press Web site at
http://www.crcpress.com

Contents

Part I Thermal Power Plant Control Process Performance and Energy Audits

Part II Thermal Power Plant Control Process Modeling

Part III Thermal Power Plant Efficiency Improvement Modeling

Part IV Thermal Power Plant Optimization Solution Supported by High-Performance Computing and Cloud Computing

Preface

The low carbon economy, environmental considerations, and fuel efficiency demands have placed strong requirements on fossil fuel–based power plants, requiring them to be operated efficiently. Improving the fossil fuel boiler combustion process is highly significant because more than 40% of the world's electricity is produced by fossil fuel, and fossil fuel power plants still play a dominant role in most countries. Even though advanced supercritical fossil fuel power generation units with carbon dioxide capture and storage (CCS) technology are utilized, some combustion-related problems like slagging and fouling often occur, decreasing boiler efficiency and increasing potential unplanned outages, and creating more concerns on regulated emissions because of the highly complex conditions changing inside the boiler.

Fossil fuel power plant boiler combustion is one of the most important processes in power generation engineering, which involves thermal dynamics, turbulent fluid flow, chemical reactions, and other complicated physical and chemical processes. Boiler combustion is a highly complex multi-input, multi-output process that is nonlinear with strong inertia. Therefore, it is difficult to establish an accurate mathematical model of boiler combustion.

Artificial intelligence (AI) technologies such as neural networks and genetic algorithms (GA) have been widely applied in the power generation industry to optimize control system processes and improve fossil fuel power plant boiler efficiency. For example, AI technology–based intelligent soot blowers are applied in coal-fired power plants to help effectively reduce slag buildup and increase the heat transfer rate of the boilers, and GA-based methods are applied to optimize fossil fuel power plant boiler combustion. However, for combustion-related problems, such as slagging and fouling, the technologies that are only dependent on AI do not work successfully because not many parameters of the boiler combustion process are measured to train the neural network–based models and acquire approximate functions for such complex processes. For example, the data regarding slagging properties are not quantified and fields of flue-gas properties are not completely measured. So AI-based boiler optimization methods are limited.

A novel method of integrating online learning, GA, and multiobjective and identification optimization with computational fluid dynamics (CFD)–based real-time simulation is proposed and developed in this research to control the fields of flue-gas properties, such as temperature and density fields, identify coal-fired power plant boiler slagging distribution, and optimize the combustion process by tuning existing proportional–integral–derivative (PID) control to improve fossil fuel power plant boiler efficiency. As compared with conventional AI-based fossil fuel boiler combustion optimization methods, the developed method in this research can obtain complete flue-gas data inside the boiler through CFD-based combustion process simulation. Moreover, the developed method in this research can not only identify slagging distribution and help soot blowers to intelligently remove the slagging, but also decrease or even avoid slagging by predictively optimizing the combustion process.

This book introduces innovative methods utilized in industrial applications, discussed in scientific research, and taught at universities. Compared with previous books published in the area of control of the power generation industry, this book focuses on how to solve highly complex industry problems regarding identification, control, and optimization through integrating conventional technologies, such as modern control technology, computational intelligence–based multiobjective identification and optimization, distributed computing, and cloud computing with CFD technology. Although the projects involved in the book just cover industry automation in electrical power engineering, the methods proposed and developed in the book can be applied in other industries such as concrete and steel production for real-time process identification, control, and optimization.

This book is divided into four parts. Part I discusses thermal power plant processes, energy conservation, and performance audits. Part II covers thermal power plant process modeling. Part III contains thermal power plant efficiency improvement modeling. Part IV discusses a thermal power plant efficiency optimization solution supported by high-performance computing integrated with cloud computing.

Part I is composed of Chapters 1 through 4. Chapter 1 introduces the equipment in a fossil fuel power plant. It also introduces combustion-related slagging and fouling, which are some of the existing difficult problems of the power generation industry, and simply analyzes how to solve the problems so as to improve fossil fuel power plant efficiency. Chapter 2 generally introduces thermal power plant processes and energy conservation, focusing on auxiliary power in power plant processes. Chapter 3 introduces energy conservation and performance audits of in-house auxiliary power equipment in a thermal power plant. Chapter 4 introduces energy conservation and performance audits of common auxiliary power equipment in a thermal power plant.

Part II contains Chapters 5 and 6. Chapter 5 discusses the processes in a fossil fuel power plant generally. The processes include energy and mass flows such as heat conduction, convection, radiation, fuel and gas flow, and water and steam flow. Deeply understanding power plant processes is significant for modeling, controlling, and improving these processes in a thermal power plant. The chapter also clarifies the main physical laws applied in power plant boiler combustion processes. Coal-fired power plant boiler combustion processes are highly complex, and heat and mass transfer are involved in these processes. Correctly choosing the exact heat and mass balance equations is important to successfully modeling, controlling, and improving these processes. Some experimental heat transfer equations are also discussed to model heat transfer processes inside the furnace of a boiler.

Chapter 6 focuses on how to develop industrial process models using MATLAB®, Simulink®, VisSim, Comsol, ANSYS, and ANSYS Fluent. Detailed model development for fossil fuel power plant boiler combustion processes is provided. Effectively using these software packages can both exactly and efficiently model, control, and optimize power plant boiler combustion processes. Chapter 6 also introduces how to develop steam turbine and generator models. It also discusses how to create a model for the integration of a boiler, turbine, and generator in a fossil fuel power plant. VisSim, MATLAB, Simulink, Comsol, ANSYS, and ANSYS Fluent are used to create models of power plant combustion processes.

Part III contains Chapters 7 through 10. Chapter 7 reviews traditional methods such as PID-based control technology and AI technology. The chapter also reviews the finite element method–supported CFD technology, which is used to simulate power plant boiler combustion. In addition, this chapter analyzes the limitation of conventional methods for the existing highly complex combustion-related slagging and fouling.

Chapter 8 clarifies how to integrate computational intelligence–based online learning with CFD technology to control temperature in a heat transfer process. The detailed method of how to integrate an online indirect adaptive controller based on the radial basis function (RBF) with CFD is given. A PID controller is also used to control the temperature. The results show that the proposed online learning integrated with CFD can control the flue-gas temperature field. In addition, the proposed method can achieve the desired objective with higher performance compared to a PID controller.

Chapter 9 covers the details of how to integrate multiobjective identification technology with CFD technology to identify the distribution of slagging inside the furnace of a coal-fired power plant boiler. A real tangential coal-fired boiler with 44 burners is simulated in three-dimensional fashion using ANSYS Fluent 14.5. The simulation achieves encouraging results compared with the corresponding results in other research. The distributed computing technology CORBA C++ is used to combine the online learning model with a CFD-based coal-fired boiler model to optimize the fields of flue-gas properties, such as flue-gas temperature and density field. In addition, digital probes are set in the model to support slagging identification. The outputs of this research show that online learning combined with CFD can identify the slagging distribution inside a coal-fired boiler.

Chapter 10 provides the innovative method of integrating computational intelligence–based multiobjective optimization with CFD to improve coal-fired power plant boiler efficiency. Two objectives are set for coal-fired boiler combustion in this research. The first objective is maintaining the coal boiler so it runs at a higher heat transfer rate. The second objective is controlling the temperature in the vicinity of the water wall tubes of the boiler and keeping the temperature within the ash melting temperature limit. Then 10 input parameters, including velocity of each burner with primary air, velocity of each burner with secondary air, primary air temperature, and secondary air temperature are adjusted to achieve the two objectives. Compared with conventional neural network–based boiler optimization methods, the method developed in the work can control and optimize the fields of flue-gas properties, such as the temperature field inside a boiler, by adjusting the temperature and velocity of primary and secondary air in coal-fired power plant boiler control systems. If the temperature in the vicinity of the water wall tubes of a boiler can be maintained within the ash melting temperature limit, then the incoming ash particles cannot melt and bond to the surface of the heat transfer equipment of a boiler and the trend of slagging inside the furnace is controlled. Furthermore, optimized boiler combustion can maintain a higher heat transfer efficiency than that of nonoptimized boiler combustion. Software is developed to realize the proposed method and obtain encouraging results through combining ANSYS 14.5, ANSYS Fluent 14.5, and CORBA C++.

Part IV contains Chapter 11, which simply focuses on how to apply this research achievement in coal-fired power plants efficiently by building an Internet-supported boiler combustion optimization platform. The chapter also analyzes the online learning and CFD-supported local boiler combustion optimization solution and Internet-based global boiler combustion optimization platform solution. In addition, how to combine high-performance computing technology, cloud computing technology, and computational intelligence–based identification, control, and optimization with CFD to build an Internet-supported industrial process optimization platform is discussed in detail. The chapter also presents the scale to which the technologies of modeling, control, and optimization discussed in the book can be extended. A list of references follows Chapter 11.

The authors of this book sincerely thank Dr. Gagandeep Singh, Jennifer Ahringer, and Kyra Lindholm of CRC Press/Taylor & Francis Group for all their help in the publication of this book. The authors also thank Professor Rajashekar P. Mandi, School of Electrical and Electronics Engineering, REVA University, Bangalore, India, and Dr. Udaykumar R. Yaragatti, Department of Electrical and Electronics Engineering, National Institute of Technology Karnataka (NITK), Surathkal, India, for contributing Chapters 2 through 4 of this book.

MATLAB® is a registered trademark of The MathWorks, Inc. For product information, please contact:

The MathWorks, Inc.
3 Apple Hill Drive
Natick, MA 01760-2098, USA
Tel: 508-647-7000
Fax: 508-647-7001
E-mail: info@mathworks.com
Web: www.mathworks.com

Xingrang Liu
University of Southern Queensland

Ramesh Bansal
University of Pretoria

Authors

Xingrang Liu, PhD, completed his doctoral degree focusing on fossil fuel power plant boiler combustion process optimization based on real-time simulation at the School of Information Technology and Electrical Engineering, University of Queensland (UQ) Brisbane, Australia, in October 2013. He completed his master study of Computer Software and Theory at Xi'an Jiaotong University in Xi'an, China, in July 2003 and undergraduate study of Computer Science and Engineering at the Northeast China Institute of Electric Power Engineering in Jilin, China, in July 1992.

He has worked as a computer engineer for 10 years and worked as a senior software engineer for 5 years in the power generation industry in China. He worked as a system developer in the Cooperative Research Centre for Integrated Engineering Asset Management, School of Engineering Systems, Queensland University of Technology (QUT), Brisbane, Australia, from 2007 to 2009. He worked as an assistant researcher and a research software engineer at UQ from 2011 to 2013. Currently, he is working as a senior software researcher at the University of Southern Queensland (USQ), Toowoomba, Austrailia.

His research interests include cloud computing and high-performance computing supported real-time control system, control system modeling, computational fluid dynamics–supported thermal power plant process modeling, and multiobjective thermal process identification and optimization.

Ramesh Bansal has more than 25 years of teaching, research, and industrial experience. Currently, he is a professor and group head (Power) in the Department of Electrical, Electronic and Computer Engineering at the University of Pretoria, South Africa. In previous postings, he was with the University of Queensland (UQ), Brisbane, Australia; University of the South Pacific, Suva, Fiji; Birla Institute of Technology and Science (BITS), Pilani, India; and All India Radio. During his sabbatical leave, he worked with Powerlink (Queensland's high voltage transmission company).

Bansal has extensive experience in the development and delivery of training programs for professional engineers. At UQ, he made significant contributions to the development and delivery of the ME Power Generation Program (a collaborative program of three of Queensland's industries and three universities). He developed and taught Generator Technology and Plant Control Systems. At BITS, he contributed to the development and delivery of the BS Power Engineering Program (for the National Thermal Power Corporation, NTPC, a 40000 MW Indian Power Generation Company) and two batches of more than 1000 students successfully completed the program.

Bansal has published over 230 research papers in journals and conferences. He has contributed several books/book chapters, including *Handbook of Renewable Energy Technology*, World Scientific Publishers, Singapore, in 2011. He has diversified research interests in the areas of renewable energy and conventional power systems, including wind, photovoltaic (PV), distributed generation, power systems analysis (reactive power/voltage control, stability, faults, and protection), smart grid, FACTS, and power quality. He is an editor of reputed journals including *IET Renewable Power Generation*, *Electric Power Components and Systems*, and *IEEE Access*. He is a fellow and chartered engineer at IET-UK, a fellow at Engineers Australia, a fellow at the Institution of Engineers (India), and a senior member at IEEE.

Part I

Thermal Power Plant Control Process Performance and Energy Audits

1

Introduction to Improving Thermal Power Plant Efficiency

1.1 Power Plant Introduction

Although good success has been achieved in power generation from renewable energy and clean energy (wind, solar, etc.), conventional fossil fuel power plants still play a significant role in both developed and developing countries. Moreover, with the development of high-steam parameter power-generation technology and carbon dioxide capturing technology, new fossil fuel power plants are becoming more efficient and environmentally friendly. A number of countries have constructed more new fossil fuel power plants because of the massive and stable unit power, low construction cost, and short construction period.

Figure 1.1 shows the boiler efficiency difference existing among various countries in the world [1]. The original design is the most important factor for boiler efficiency, for example, the efficiency of ultrasupercritical boilers is much higher than conventional subcritical boilers. However, the operation technology plays a significant role in improving boiler efficiency [2]. Boilers of identical design apparently firing identical fuels have often been reported to encounter quite different slagging and fouling problems [3]. This is the main reason why a number of famous international companies, such as ABB, Bailey, and Honeywell, provide their commercial software tools to optimize boiler efficiency in the power-generation industry [4]. In addition, almost every power-generation company in the world is making efforts to increase boiler efficiency and limit negative environmental emissions.

Artificial intelligence–based methods have been applied in a number of power stations to optimize the combustion process and increase efficiency [5–9]. In addition, computational fluid dynamics (CFD) is used to model the complex combustion process, achieving successful assessment of boiler performance [10–15]. However, some combustion process-related problems, such as slagging and fouling, are still plaguing the electricity-generation industry by impairing boiler efficiency and increasing unplanned downtime caused by slagging blockage in critical components [16]. The main reasons

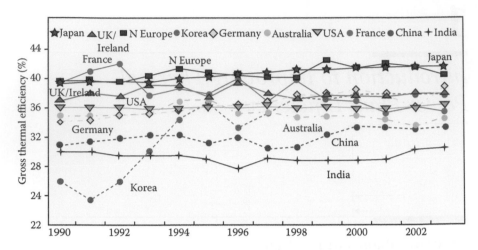

FIGURE 1.1

Comparison of gross thermal efficiency (low heat value base) of coal-fired power plants. (*Source:* Updated comparison of power efficiency on grid level, ECOFYS Co., 2006.)

for slagging and fouling are found to be both boiler design and operation. Moreover, with fuel quality frequently changing and critical boiler equipment gradually degrading, how to control an optimal flue gas and limit unburned carbon are significant to improving boiler efficiency [17].

1.2 Specific Problems of Fossil Fuel Boiler Combustion

Systematic analysis of the coal-fired boiler-combustion process uncovers three main influences of boiler efficiency. First, high-level slagging and fouling can massively reduce the heat-transfer rate of the water wall, superheater, or other heat-transfer equipment. Limiting slagging and fouling build-up can improve the combustion efficiency. Second, if the temperature or volume of exit gas is too high, more heat will be wasted by exit gas. Therefore, controlling the temperature and volume of exit gas can help achieve high efficiency. Finally, an unburnt gas or solid carbon is another negative factor to boiler efficiency. Decreasing unburnt carbon can effectively improve boiler combustion efficiency.

Figure 1.2 shows that fuel with specific characteristics is sent to the mill where the coal is pulverized and blown into the furnace of a boiler from burners by mixing with the primary air. Measurement point 1 can measure the amount of coal which is sent to mill, and point 2 can measure the amount of primary air which is mixed with coal powder. The coal-powder concentration can be measured at point 3. The flow speed of the mixture of air and coal powder can be measured at point 4. The temperature of the mixture of coal and air can be measured at point 5. The excess air rate can be measured

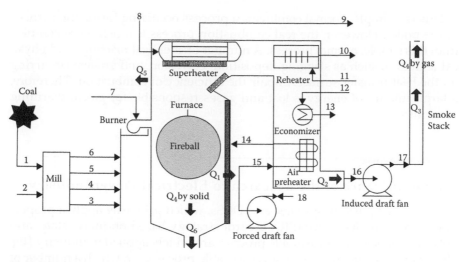

FIGURE 1.2
Heat-flux distribution in the furnace of a coal-fired power plant boiler while combustion process is running.

at point 6. All the boiler input parameters can be adjusted and tuned in the control system, such as a programmable logic controller (PLC)-based control loop or distributed control system (DCS) in a power plant.

The burners are installed in the wall of furnace and the mixture of fuel and air is blown into the furnace to burn from burners. The secondary air is applied to adjust the flame shape of the fireball. It can be measured at point 7. With appropriate adjustment of the angle of burners, a rotated fireball can be formed in the furnace of the boiler. A high percentage of heat radiates to the surface of the water wall and the superheaters. At the same time, heat conduction occurs on the heat-transfer surface of all equipment inside the furnace. As the flow in the convection passes of furnace, the flue-gas transports the residual heat outside the boiler. The overall process is shown in Figure 1.2. The saturated steam, which is heated in the boiler, drives the turbine with high enthalpy because of its high temperature and pressure. Measurement points 8 and 9 can measure the characteristics of the saturated steam in the primary and secondary superheaters. The reheater, economizer, and air preheater are installed in the flue gas pass to recover the residual heat. The temperature and pressure of the steam in the reheater can be measured at points 10 and 11. The temperature of feed water in the economizer is measured at points 12 and 13. The temperature of the primary air inside the reheater can be measured at points 14 and 15.

The residual flue gas blows out from the smoke stack and the temperature and pressure of the exit gas can be measured at point 16. A forced draft (FD) fan and induced draft (ID) fan keep a correct draft inside of the furnace. The power of the fans can be measured at points 17 and 18.

This is a simple, normal combustion process occurring inside the furnace of the boiler. However, the real combustion process is much more complex than this simple, normal process. A number of chemical reactions and physical activities, such as slagging deposition, corrosion, and erosion occurring on the heat-transfer surface, impair the efficiency of combustion. Therefore, a large amount of energy is lost and gases responsible for global warming are emitted.

1.3 Significance of the Research to Electrical Power Industry

PLCs, data acquisition systems, and DCSs, which primarily apply a proportional–integral–derivative (PID) control strategy based on input, state, and output variables in a measurable process, are widely applied in industry [18]. However, there are also some immeasurable processes in which a number of critical parameters are impossible to measure. For example, some parameters including slag thickness, slag accumulation, and corrosion rate are difficult to read using traditional instruments from the boiler combustion process, which is a highly complex and significant process in a power plant because of existing states of equipment or work fluid physical or chemical properties.

Statistics from Figure 1.1 show that there is much difference in fossil fuel boiler efficiency among various countries in the world. Improving the fossil fuel boiler efficiency and reducing carbon dioxide emissions are significant for countries with not only high average boiler efficiency but also low average boiler efficiency. The fossil fuel boiler combustion process is a highly complex dynamic system. As boiler conditions such as variation of fuel quality and equipment age change, the boiler is not able to maintain the satisfactory status of its original design. In this case, the widely applied conventional PID controllers become insufficient to effectively control the performance. For example, set-point values originally set for some input parameters of a controller may not be the optimum option after coal quality is changed or the furnace becomes dirty.

Moreover, the neural network–based advanced controlling approach does not always work successfully. For example, slagging and fouling accumulated on the surface of heat-transfer equipment or the heat convection pass not only impair boiler combustion efficiency but also lead to potential severe threats to the boiler. It is difficult to restrict slagging and fouling increase due to nonaccurate reading of data on slagging and fouling status. With system data from the instrument, the neural network can be trained to approximate highly nonlinear functions, since the neural network depends on the input/output data but not on the physical structure of the system. It is flexible and can be easily adapted to different types of power plants. However, it does not work without instrument data for training.

In this solution, neural network–based online learning technology integrated with real-time numerical simulation based on CFD technology is proposed to solve the problems severely impairing combustion process efficiency. Moreover, the proposed strategy is extended to a global fossil fuel boiler combustion monitoring, evaluation, and tuning platform based on web services, a large-relation database, and multiagent and high-performance-computing technology. The platform aims to provide real-time fossil fuel boiler combustion condition monitoring, evaluation, and real-time combustion control tuning through which all global boilers can assess each other and obtain optimum control-input parameters to improve combustion efficiency.

The proposed solution in this work tries to limit slagging accumulation in the water wall and superheater and maintain an optimal cleanness of the heat-transfer surface to achieve a maximum of Q1 and obtain optimal heat-transfer efficiency. In addition, effective methods are applied in the proposed solution to decrease the exit gas heat loss Q2, unburned combustible gas loss Q3, and unburned carbon heat loss Q4.

Figure 1.3 shows the structure of the proposed platform and the local power plant computer system structure. The proposed platform includes four data layers: online multiobject and optimization, multiagent, distributed computing, and distributed resources. The distributed data and high-performance vector machines can be shared globally by applying cloud computing, web services, and distributed technologies. The proposed methods that integrate online learning technology with real-time simulation are implemented on the platform. Moreover, the proposed methods are extended from a local power plant to global power plants by using the proposed platform, which is supported by high-performance computing, cloud computing, web services, and multiagent technologies. Depending on the platform and online learning technology applied on the platform, power plant boilers can assess each other. In addition, multiobject-optimization technology is used to tune the local power plant boilers online.

Figure 1.3 also shows the relationship between the platform data and the local computer system data. After an identification or optimization, a request is accepted by the remote agent running on the proposed platform. The real-time dynamic data and the required stationary data of the boiler are transferred to the platform and processed by high performance computing based on a specific identification or optimization model. The result will feed back to local power plants to tune their controllers using the local optimization system. Figure 1.5 shows that the local optimization system is supported by DCS and other computer control systems. At the same time, the optimization system can monitor and control the local computer control systems.

Integrating real-time simulation with online learning technology is applied in this research to build a slagging level and distribution-identification model, fuel-quality-identification model, and combustion fireball-control model to improve both conventional and advanced boiler

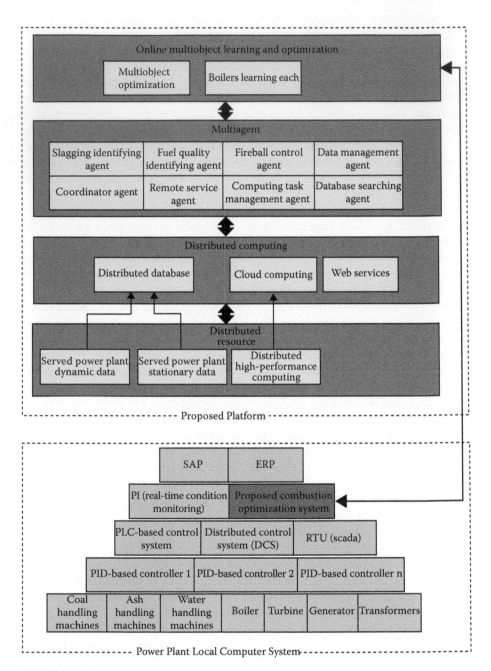

FIGURE 1.3
Structure of the proposed platform and a local power plant computer system structure.

efficiency. In addition, Internet technology, distributed computing, and a high-performance technology-supported platform are used to build up a platform on which the boiler slagging level and distribution-identification model, coal-quality model, and combustion fireball-control model run. This platform can provide online monitoring and tuning service to all boilers distributed in different places in the world. Moreover, boilers with normal efficiency can assess using their corresponding boilers with high efficiency and become more efficient based on this platform.

1.4 Fouling and Slagging Distribution-Identification Model

Online learning technology integrated with CFD technology will be applied to build a fireball model, water-wall slagging-formation model, furnace flue-gas model, and superheater fouling-accumulation model in which high-performance computing, multiagent technology, and fast finite-element-method technology will be applied to improve the computing in CFD. In addition, these models are run on the multiagent and high-performance computing-based platform that can connect boilers in different places to do remote online condition monitoring, evaluation, and tuning. Therefore, local models that run on the local power plant and central models that run on the remote platform will be created.

The fuel quality, pulverizera, furnace instrument, exit flue gas, feed water and saturated steam, FD and ID fan, and power data will be collected to build the local models. Figure 1.4 shows the heat flow balance relation and how to partition the furnace and gas flue convection area while using finite element–based CFD to create models. The furnace area and gas flue convection area will be partitioned in different subareas. Then the heat rate difference between the clean subarea and current subarea in the corresponding water or steam side will be obtained. Based on these data, a boiler furnace slagging distribution diagram can be obtained to show the current slagging situation of the boiler. Technology allowing boilers to learn from each other will be applied to build these models based on cloud computing, web services, high-performance computing, multiagent systems, and a distributed database. A standard slagging and fouling level based on the standard fuel quality level and the slagging influence on the heat-transfer efficiency will be created in the solution.

Figure 1.5 shows the logic of the slagging and fouling identification model which will be implemented on the platform. The online learning module adjusts the weight matrix of the neural network to identify the most matchable data for input of the real-time simulation module, which applies CFD, the compressible Navier–Stokes equation, and a large eddy simulation model to simulate the local boiler combustion process by obtaining dynamic geometry boundary condition data.

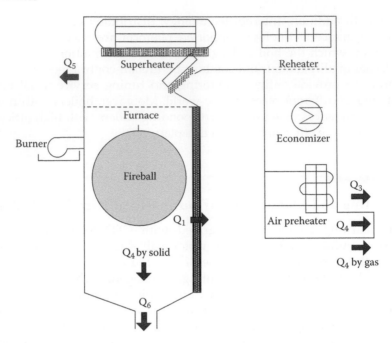

FIGURE 1.4
Method to identify the slagging level and distribution inside the furnace of a coal-fired power plant boiler.

1.5 Fireball Control and Optimization Model

The standard fuel-quality class will be created based on the heat rate and potential slagging formation level. First, a CFD-based fireball model will be created. Then movement of the fireball in the model will be designed by adjusting the fireball position. Finally, the difference of the heat rate of the model output between the two different positions will be compared and the standard level of the current slagging impact rate in the model will be obtained. Based on the standard slagging impact rate data, the standard fuel quality of the current fuel can be obtained. Figure 1.6 shows adjustment of the fireball to obtain the current coal quality.

Online learning technology integrated with CFD technology will be applied to create the model based on the proposed platform, supported by cloud computing, a distributed database, web services, and multiagent and high-performance-computing technology. Boilers can assess each other and the coal quality identifying model can be created. Figure 1.7 shows the logic of the coal quality identification model which will be implemented on the platform.

The fireball model will be created to maintain the appropriate exit flue gas temperature and minimize unburnt gas and solid carbon. Figure 1.8

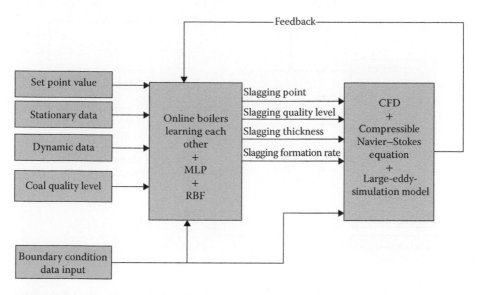

FIGURE 1.5
Logic of how to identify the distribution of slagging and fouling inside a furnace based on computation intelligence integrating with CFD.

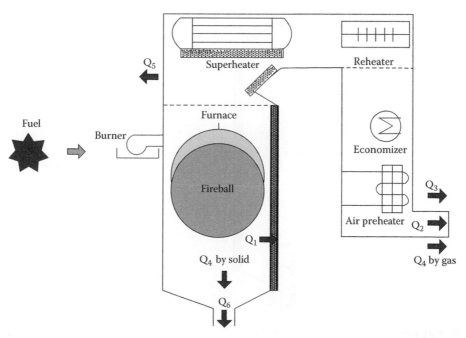

FIGURE 1.6
Mechanism of how to identify coal quality based on CFD technology.

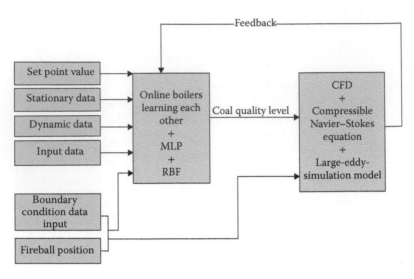

FIGURE 1.7
Logic of how to identify coal quality based on computational intelligence integrating with CFD.

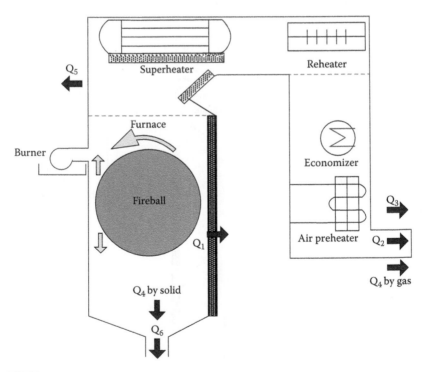

FIGURE 1.8
Strategy of how to control the fireball with optimal position and rotating speed in the furnace of a coal-fired power plant boiler.

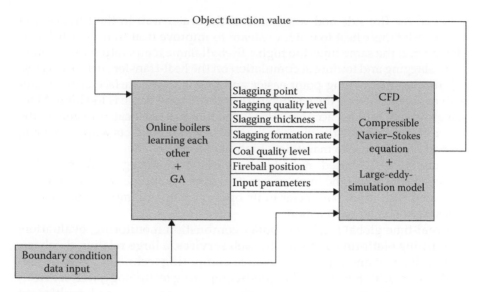

FIGURE 1.9
Logic of how to control fireball to improve coal-fired power plant boiler combustion process.

shows the rotating speed and the height of the fireball that will be adjusted to maintain an optimal fireball position by tuning furnace-input parameters such as coal fineness, speed of mixture of coal and air, speed and amount of second air, excess air, speed and amount of ID and FD fan, and the burner-tilting angle. Figure 1.9 shows the logic of the fireball control model. The online learning module includes online learning and a genetic algorithm (GA). Based on the proposed platform supported by cloud computing, a distributed database, web services, and multiagent and high-performance-computing technology, boilers can assess each other and use GA to find the optimal inputs to tune the local boiler combustion process online.

1.6 Slagging Distribution Identification and Combustion Optimization

An optimization platform supported by Internet-based technologies and high-performance computing for boiler combustion is a highly complex process in which parameters of each subprocess are related and the multiobjects of these subprocesses may conflict with each other. For example, maintaining a high temperature of the fireball flame increases the temperature difference between the fire side and water side of the water-wall in the furnace and can transfer more heat to water or steam to improve heat-transfer efficiency. However, at the same time, too high a fireball flame temperature may lead to more slagging and fouling accumulation on the heat-transfer surface because ash reaches the melting point easier and bonds to the surface of the water wall or superheaters if the temperature of the fireball flame is too high. More slagging and fouling accumulation seriously restrict heat transfer to the water or steam side of the pipes of the boiler, which conflicts with increasing fireball temperature to maintain a high-heat-transfer rate.

Therefore, a GA-based multiobject optimization strategy will be applied to control and adjust the input parameters of each subprocess of combustion and keep each input parameter in its optimal status to improve the overall efficiency of boiler combustion.

A real-time global fossil fuel boiler combustion monitoring, evaluation, and tuning platform supported by web services, a large relation database, and multiagent and high-performance-computing technology will be created. Figure 1.10 shows how the cloud-computing technology, web services, distributed database, and high-performance-computing and multiagent technologies collect the distributed high-performance vector machines to build a platform that can provide online identification and tune the conventional controller in global power plants. In addition, the figure shows that it is very easy for a local power plant to request real-time tuning to improve the efficiency of their power plant. The only requirement will be to set up an industrial personal computer and install a local optimization system. The platform makes full use of vector machines distributed in different universities that provide remote service for power plants in different parts of the world to help monitor, evaluate, and tune the combustion process of a boiler.

First, web services and a distributed database are applied to create the platform in which different boilers can be connected using the Internet and all kinds of boiler data will be accumulated to support the boilers in assessing each other based on the platform. Second, slagging, fuel quality character, fireball control, computing task management, database searching, and remote service agents are created to form the boiler optimization platform, in which boilers assess each other using web services, databases, and online learning technology integrated with CFD technology, and multiagent and high-performance-computing technologies. The computing task management agent can obtain the high-performance computing from one vector machine or a cluster of vector machines distributed in different places. Finally, OVERset grid FLOW solver (OVERFLOW) codes will be developed to speed up all the models on the platform. Moreover, the solution will apply fast finite-element-algorithm technology to improve the speed of the model computing and also provide a local boiler combustion optimization system. Figure 1.11 shows the local optimizing system and platform-based boiler combustion monitoring, evaluating, and tuning system.

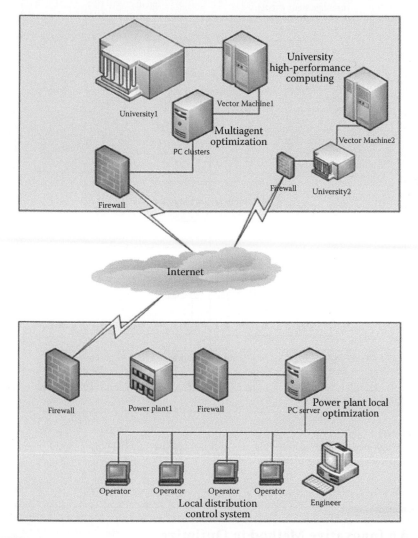

FIGURE 1.10
A proposed platform supported by cloud computing, web services, multiagent, high-performance computing, and a distributed database providing condition monitoring, evaluation, and tuning to global boilers.

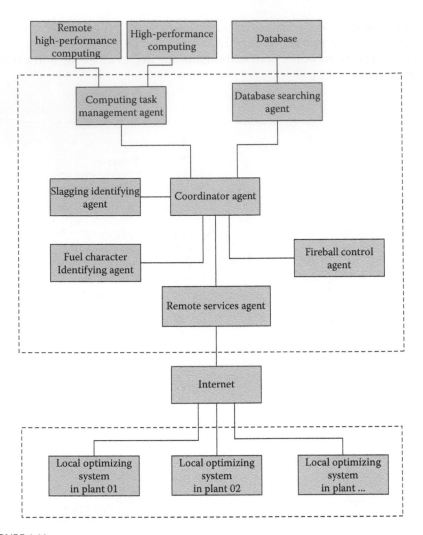

FIGURE 1.11
Logic of high-performance computing and multiagent-based global boiler condition monitoring, evaluation, and tuning platform.

1.7 An Innovative Method to Optimize Fossil Fuel Power Plant Combustion and Limiting or Even Removing the Tendency of Slagging

The existing and widely adopted soot blowers can only frequently try to remove slagging and fouling; they cannot prevent slagging. As both computer software and hardware technologies have advanced dramatically and

CFD technology can be used to simulate highly complex processes such as boiler combustion with accurate results [18–23], this research proposes a novel method to solve combustion-related problems such as slagging.

This research integrates CFD with online learning, GAs, and multiobjective optimization to achieve an improved combustion process and reduce or even prevent the tendency of slag buildup while coal quality and gas fluid fields are frequently changing by optimizing the fields of temperature and velocity of primary and secondary air. The optimal field inside the furnace of the boiler can effectively increase boiler efficiency, reduce the trend of slagging, and decrease the carbon emission.

1.8 Creating a Novel Method to Identify the Distribution of Slagging inside of a Coal-Fired Boiler

This research has provided a method to identify slagging distribution and quantify slagging and fouling inside a boiler. The results can be fed back into the distributed control system to keep the combustion process in an appropriate state. In addition, the method can not only help to make conventional soot blowers operate intelligently but also monitor slagging and avoid equipment damage caused by serious slagging and fouling. Furthermore, the method can be applied to support predictive maintenance in coal-fired power plants.

1.9 Conclusions

Thermal power plant processes, thermal power plant efficiency problems, and the solutions achieved during this research are discussed in this chapter. The detailed processes of a power plant and energy conservation for the main thermal power plant processes are discussed in the later chapters. In addition, the methods currently used in some thermal power plants for energy conservation and performance audits are discussed in more detail in Chapters 2 through 4. Furthermore, new methods of modeling, controlling, and improving thermal power plants efficiency are clarified in Chapters 5 through 10.

2

Overview of Energy Conservation of Auxiliary Power in Power Plant Processes

Rajashekar P. Mandi

School of Electrical and Electronics Engineering, REVA University, Bangalore, India

Udaykumar R. Yaragatti

Department of Electrical and Electronics Engineering, National Institute of Technology Karnataka (NITK), Surathkal, India

2.1 Introduction

The major portion of electrical power is produced from combustion of fossil fuels, especially coal [24]. Electrical power generation through combustion of fossil fuels like coal emits gases like carbon dioxide (CO_2), sulfur oxides (SO_x), nitrogen oxides (NO_x), chlorofluorocarbons (CFC), and suspended particulate matter (SPM). Global warming is a great concern due to the burning of fossil fuel for electrical power generation in thermal power plants. World electrical energy increased from 10,122 to 20,353 TWh/year from 1990 to 2014 and is shown in Figure 2.1 [25]. CO_2 emission from electrical power generation is about 13,393 million t/year (41.2% of total CO_2 emission). Worldwide electrical energy generation through CO_2-emitting technology (fossil fuel based) is about 86.9% of total electrical energy generation and non–CO_2 emitting technology is 13.1% of total electrical energy generation (see Figure 2.2) [25]. The energy share for renewable is about 1.9%, hydro is about 6.7%, and nuclear is about 4.5%. This shows that major electrical energy comes from fossil fuel–based power plants, which release more pollutants into the atmosphere.

Electrical energy generation in India increased from 212 (2.09% of world energy) to 911 TWh/year (4.49% of world energy) from 1990 to 2014 as shown in Figure 2.3 [25]. Indian electrical energy generation through CO_2-emitting technology (fossil fuel based) is about 69.6% of total electrical energy generation and non–CO_2 emitting technology is 30.4% of total electrical energy generation (see Figure 2.4). The energy share of renewable is about 13.0%,

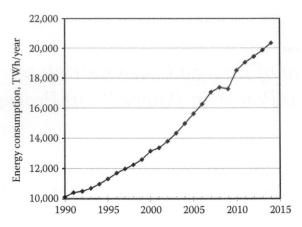

FIGURE 2.1
World electrical energy consumption.

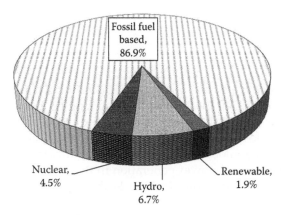

FIGURE 2.2
Share of world electrical energy generation.

hydro is about 15.3%, and nuclear is about 2.1%. Energy generation in India through non–CO_2 emitting routes is better than the world's average. Since the major portion of electrical power comes from CO_2-emitting power-generating technology, enhancing the energy efficiency of electrical power generation of conventional fossil fuel–based plants and energy conservation plays a major role in controlling global warming as well as reduction of greenhouse gases.

The net coal to electrical power (grid) conversion efficiency of the coal-fired thermal power stations in India varies between 19.2% (30 MW unit) and 30.7% (500 MW unit). The auxiliary power (AP) used for coal-fired stations varies with the size of the power plants, that is, it varies between 5.2% (500 MW unit) and 12.3% (30 MW unit) at 100% maximum continuous rating

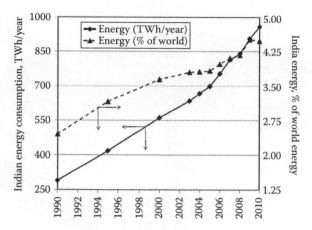

FIGURE 2.3
Indian electrical energy consumption.

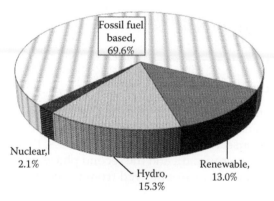

FIGURE 2.4
Share of Indian electrical energy generation.

(MCR). The computed AP used by the coal-fired power plants in India is about 11,340 MW, which will be about 8.4% of power generation by coal-fired power plants. The AP is on the higher side in Indian thermal power plants compared with other developed countries due to poor equipment performance, the use of suboptimal coal quality, excessive steam and water flow, internal leakage in equipment, inefficient and obsolete drives, aging of equipment without proper maintenance, design constraints like oversizing of equipment, and so on.

Figure 2.5 shows the share of power generation by different types of power plants in India as of March 31, 2014 [26]. The major electrical power comes from 210 MW units (26.2% of total installed capacity) followed by 500 MW units (26.0% of total installed capacity).

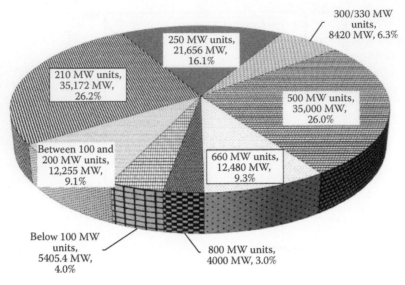

FIGURE 2.5
Power share by different units in India.

2.2 Energy Conservation

Energy conservation is a technique to save and optimize energy consumption through enhancing the energy efficiency of equipment and processes. Basically, energy conservation consists of two phases, that is, the first phase is the conduct of an energy audit and the second phase is the implementation of energy conservation measures derived from the energy audit work.

2.2.1 Energy Audit

An energy audit is a technique used to identify the pattern of energy consumption, and get information on the present operating efficiencies of equipment, identifying how and where energy losses are occurring in the system, analyzing the system, and providing feasible solutions to conserve energy with a detailed techno-economic evaluation. These energy conservation measures lead to a plan for renovation and modernization (R and M) of equipment.

An energy audit consists of mainly three phases:

1. Preliminary energy audit
2. Detailed energy audit
3. Report preparation

2.2.1.1 Preliminary Energy Audit

A preliminary energy audit pertains to a preliminary survey of the energy flows and process at the plant to get a holistic picture of the plant and its energy efficiency. A preliminary audit is carried out in a limited span of time and is sometimes called a walk-through audit. It focuses on the major energy supplies and demands, accounting for at least 70% of the total energy requirement. The energy auditor examines the data already available in the plant. These data include

- Layout of the electrical distribution system
- Energy flow diagram
- Nameplate details for major energy-consuming equipment
- Manufacturers catalog for major energy-consuming equipment
- Log sheet data on energy consumption

For the purpose of a preliminary energy audit, a questionnaire is prepared to extract preliminary information concerning the site, its functions, and the activities being conducted. After collection and analysis of the energy use and cost data, the next stage is to examine the ways by which a detailed audit can be conducted in the plant.

2.2.1.2 Detailed Energy Audit

The detailed energy audit includes measurement at the site and performance evaluation of all individual equipment:

1. Instrumentation: Measurements are important aspects of an energy audit and are essential during the detailed audit. It may be required to calibrate all the instruments and obtain a set of readings to correct the difference in the measured quantities. Instrumentation measurements may be taken from online instruments if facilities for instrumentation are available or with portable or fixed instruments carried by the energy auditors.

2. Data analysis: The data analysis includes evaluation of the performance indices such as efficiency, specific energy consumption (SEC), specific energy generation (SEG), specific fuel consumption (SFC), load factors, specific loss, and so on. The data collected are analyzed. The analysis will lead to computation of indices, which form the basis for identifying

   - The areas of energy loss
   - Possibility of eliminating or modifying production processes
   - Possibility of waste heat/energy recovery

It also helps in computing

- Comparison of the plant performance with the design parameters (i.e., predicted performance parameters)
- Comparison of the deviations between design and operating parameters
- Evaluation of proposals for energy efficiency measures
- Feasibility study of recommended energy efficiency measures

3. Identification of energy efficiency measures: The actual computed data will be compared with the predicted performance data given by the manufacturer. The present condition of the equipment, variation in input parameters, age factor of equipment, weather conditions, and so on will be considered. If the performance parameters are deviating widely from the design conditions, the different energy efficiency schemes will be derived with various options.

4. Techno-economic analysis: The next step is identifying the areas where energy efficiency is economically effective and reliable. The energy savings for each piece of equipment will be evaluated. The total cost of energy efficiency measures and annual savings should also be calculated. Suggestions should be provided to implement the energy efficiency techniques and energy efficient equipment. The payback period and returns on investment are computed to assess the actual economics. The feasibility and viability study will also be carried out for the various energy efficiency schemes. The payback period will be calculated by considering various factors like net present value (NPV) of the existing equipment, interest rate, inflation rates, overhead cost, installation charges, running charges, and so on.

5. Energy efficiency suggestions: The suggestions will be presented in three sections, that is, immediate measures, medium-term measures, and long-term measures. **Immediate measures** require very little direct investment, the payback period is less, and technology availability is easy. **Medium-term measures** include slightly higher investment, minimum gestation period, and a moderate payback period. **Long-term measures** are comprised of heavy investment, a high gestation period, renovation and modernization, a prolonged payback period, technological development, and so on. The simple payback period will be estimated while considering the interest and other operational and maintenance costs on the capital. The list of suppliers and product catalogs of various energy-conserving devices will also be provided at the time of providing recommendations for energy conservation measures.

6. Quantification of energy savings: The overall savings from immediate, medium-, and long-term measures will be quantified. The

savings index will be evolved by considering improvement in plant performance, energy and fuel savings, investment, and so on.

2.2.1.3 Energy Audit Report

The draft report will be prepared and submitted to plant officials after the quantification of fuel and energy savings, and computation of the payback period for individual suggestions. The draft report will contain methodology, scope of work, results of efficiency tests, performance indices, and a proposal for energy efficiency improvement schemes with details of the investment and payback period. The draft report will be presented. These suggestions are then discussed with plant engineers and operators.

After receiving comments from the plants, a detailed discussion with the plant officials will be held. Each suggestion will be looked at for technological availability, techno-economic viabilities, user-friendly scheme for generation and maintenance, and so on. Then a final report will be prepared including the mutually agreed viable energy efficiency measures.

2.3 Auxiliary Power

The AP in a thermal power plant is the power used to drive the auxiliary equipment required to start and run the power plants. The AP is also called the works power, parasitic power, or in-house power. It is expressed in units of power (kW or MW) or as a percentage of gross generated power.

The energy performance parameters like AP, SFC, specific oil consumption, heat rate (turbine efficiency), boiler efficiency, generator efficiency, and so on depends on the plant load factor (PLF). Again the PLF depends on the fuel quality, availability of resources, forced outages, grid conditions, performance of individual equipment, and so on. The PLF of thermal power plants varies widely for different plant capacities. To achieve better performance for a plant, the unit must run at a higher PLF. The progressive changes in technologies and upgrade of unit sizes in new power plants improved the average PLF of Indian coal-fired power plants from 52.4% (1985–1986) to 78.6% (2007–2008); this slightly decreased to 70% (2012–2013) due to nonstabilized units (i.e., introduction of low-cost and inferior-quality units) (see Figure 2.6). The AP reduced from 9.76% (1992–1993) to 8.17% (2007–2008) and again slightly increased to 8.44% during 2011–2012. Figure 2.7 shows the variation of AP with annual average PLF and as the PLF increases the AP is decreased [27].

The average AP of coal-fired thermal power plants for developed countries (subcritical technology with motor-driven boiler feed pump [BFP]) is about

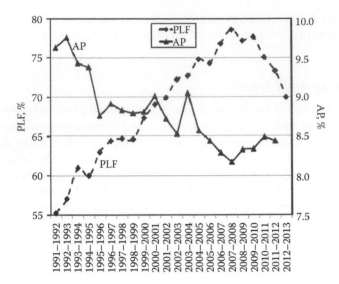

FIGURE 2.6
Variation of plant load factor (PLF) and auxiliary power (AP) in India.

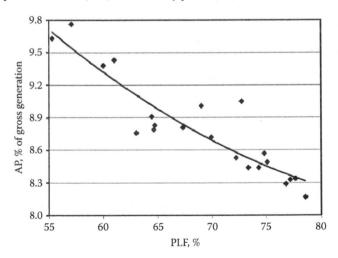

FIGURE 2.7
Variation of AP with PLF.

7.6% [28] and is lower than Indian power plants due to the use of low ash and higher calorific value coal. As per the report on "compendium best practices for coal-based power plants in Germany" prepared by VGB Power Tech. (Essen, Germany), the AP of 600 MW units in Germany is 7.4% [29]. The normative AP prescribed by regulatory norms in India is given in Table 2.1 [30].

All the aforementioned AP figures are for power plants operating above 80% PLF up to MCR condition. The MCR of a generating unit means

TABLE 2.1

Auxiliary Power (AP) as per Regulatory Norms in India

Sl. No.	Unit Capacity	AP, % With Natural Draft CL or without CT	With FD CT
01	200 MW plant series	8.5	9.0
02	500 MW and above with steam-driven BFPs	6.0	6.5
03	500 MW and above with motor-driven BFPs	8.5	9.0

Note: CT, cooling tower; FD, forced draft; BFP, boiler feed pump.

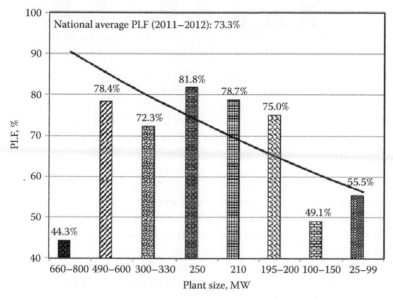

FIGURE 2.8
Average PLF of different units.

operating the plant at full load megawatt output capacity of a unit which can work continuously at specified conditions.

Figure 2.8 shows the average PLF of different units and Figure 2.9 shows the average AP for different units during 2011–2012 in India. The PLF of 660–800 MW units were not stabilized and PLF was about 44.3%, much less than the national average value of 73.3%. The average PLF of 800 MW units at Ultra Mega Power Projects (UMPP) is 91.8%. The average AP at UMPP is 8.1% and is lower than the national average value but is higher than 500 MW (i.e., about 5.2%–6.2%), because BFPs are motor driven for 800 MW units compared with the steam-driven BFPs used for 500 MW units. The average PLF of 195–600 MW units is in the range of 72.3%–81.8% and is normal, whereas the average PLF of 25–150 MW units is on the lower side in the range of 49.1%–55.5% due to the use of very old technology, obsolete equipment, aging

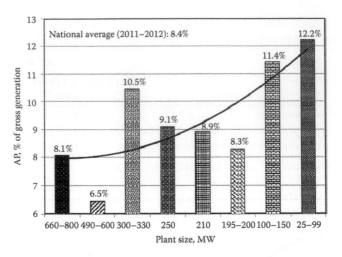

FIGURE 2.9
Average AP of different units.

factors, manual operation, and so on. The lower PLF of units increases the AP, which is in the range of 11.4%–12.2%. The average PLF of 250 MW units is 81.8%, better than the national average value of 73.3%. Even some of the 250 MW power plants like Dahanu TPS are working at 100.5% PLF and Chabra TPS units are working at 91.6%. The average AP of the 250 MW group is 9.1%, which is higher compared with national average value of 8.4%, because at Dahanu TPS about 1% of additional AP is used for a flue-gas desulfurization (FGD) plant to control the environmental pollution due to stringent pollution norms. The AP of 300–330 MW units is on the higher side at about 10.5% compared with 195–250 MW units whose AP is in the range of 8.3%–9.1% because of the inferior quality of equipment used for 300 MW units.

Higher AP is mainly due to

1. Fuel quality
   - Use of poor coal quality like high ash content, low calorific value (CV), and so on
   - Scarcity in supply of fuel
   - Poor linkage of coal
   - Nonoptimal raw coal at mill inlet and pulverized coal at mill outlet
2. Operational optimization
   - Lack of coordination and operational optimization
   - Higher turbine heat rate (excessive steam flow), that is, higher specific steam consumption
   - Lower boiler efficiency
   - Higher de-mineralized (DM) water makeup

3. Plant maintenance
   - Internal leakage
   - Prolonged operation of plant without periodic annual overhaul
   - Operation of plant with steam leakages due to nonavailability of unit for taking up for maintenance
4. Design constraints, technological advances, and retrofitting
   - Inefficient drives
   - Oversizing of equipment
   - Use of very old and obsolete controls
   - Aging of equipment
   - Obsolete equipment

The AP is broadly categorized into in-house AP and common (outlying) AP (Figure 2.10). The in-house AP is the power that is essential and directly related to the individual units, whereas the common (outlying) AP is the power used to run the common auxiliaries and other outlying equipment.

The power supply to these in-house auxiliaries is fed from unit auxiliary transformers (UATs). UATs will step down the voltage at the generator terminal of 15.75 kV (210 MW plant) to 6.6 kV (UAT bus voltage). Generally for better availability and reliability, two UATs with two UAT buses or one UAT with two UAT buses (i.e., UAT "A" and UAT "B") are provided. Each UAT bus provides the power supply to half of the total in-house high-tension (HT) equipment. Figure 2.11 shows the schematic of a single-line diagram of the AP distribution system in a typical 210 MW power plant. During startup of the unit, the power to these UATs is provided from station transformers (STs) by operating the tie between the UAT bus and ST bus.

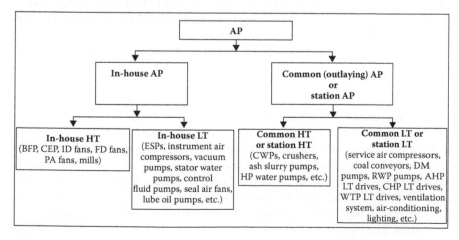

FIGURE 2.10
Schematic of AP in a typical power plant.

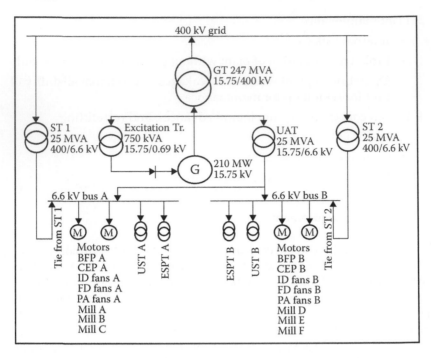

FIGURE 2.11
Schematic of single-line diagram of AP distribution.

The turbine auxiliary system, such as pumps like BFPs, condensate extraction pumps (CEP), and so on, which accounts for about 52%–58% of the total AP, is followed by the boiler auxiliary equipment, such as fans like induced draft (ID) fans, forced draft (FD) fans, primary air (PA) fans, mills, and so on, which accounts for about 30%–35% of the total AP for a typical 210 MW plants [31].

The AP is measured at 100% PLF in a typical 210 MW plant and the results are given in Table 2.2.

The measured in-house HT AP is in the range of 7.66%–9.65% of gross generation (for the plant load at MCR condition and 70% PLF). The excitation power used for a static-excited generator is in the range of 0.25%–0.30% and generator transformer (GT) losses are in the range of 0.28%–0.35% of gross energy generation. The measured in-house LT AP is in the range of 1.52%–2.08% and the total measured in-house AP is in the range of 7.66%–9.65%. The measured common AP is in the range of 1.63%–2.29%. The measured overall AP of a typical 210 MW power plant is in the range of 9.29%–11.94%, which is on the higher side compared with national and international average values. The optimum values (for plant operating at MCR condition) are derived from performance tests conducted on more than 37 units of similar plants in India.

TABLE 2.2

Auxiliary Power (AP) of a Typical 210 MW Plant

Sl. No.	Particulars	Measured Range of AP (70% PLF to MCR), % of Gross Generation	Optimal Value, % of Gross Generation
01	BFPs	2.44–2.96	2.25
02	Condensate extraction pumps	0.22–0.30	0.20
03	ID fans	1.12–1.32	0.80
04	FD fans	0.23–0.28	0.24
05	PA fans	0.94–1.20	0.70
06	Mills	0.66–0.86	0.50
07	In-house HT AP	5.61–6.92	4.69
08	In-house LT and losses	1.52–2.08	1.15
09	Generator excitation power (static)	0.25–0.30	0.25
10	GT losses	0.28–0.35	0.28
11	Total in-house AP	7.66–9.65	6.37
12	CHP	0.12–0.22	0.10
13	AHP	0.10–0.25	0.07
14	WTP and RWP	0.10–0.16	0.06
15	Air compressor	0.07–0.09	0.07
16	CWP with CT and GSP/ACW	1.20–1.90	1.33
17	Common AP	1.63–2.29	1.63
18	Total AP	9.29–11.94	8.00

Note: PLF, plant load factor; MCR, maximum continuous rating; ID, induced draft; PA, primary air; HT, high tension; LT, low tension; GT, generator transformer; CHP, coal handling plant; AHP, ash handling plant; WTP, water treatment plant; RWP, raw water pumps; CWP, circulating water pumps; ACW, auxiliary cooling pump.

2.3.1 Total AP

The total AP or station AP is the net power difference between the gross power generation (measured at generator terminals) by all the units in the plant and power export to the grid (i.e., GT output minus station transformer import) or the net power sent to the grid through outgoing feeders or lines. The station AP (MW) is computed by

$$P_{station} = \frac{1}{720} \times \left[\sum_{i=1}^{i=n} (E_G)_i - \left(\sum_{k=1}^{k=m} (E_F)_k - \sum_{j=1}^{j=p} (E_{ST})_j \right) \right] \quad (2.1)$$

where E_G is the gross energy generation by the individual unit (MWh/month), i is the number of generators in the station (i varies from 1 to n units), E_F is the energy sent (export) by the individual feeder into the grid (MWh/month), k is the number of outgoing feeders in the station (k varies from 1 to

m feeders), E_{ST} is the energy consumption by the ST (MWh/month), and j is the number of ST in the station (j varies from 1 to p ST).

Generally, the AP is computed as percentage of gross generation, which is the ratio of power input in magnitude to gross power generated at generator terminals and is computed as

$$AP_{station} = \frac{P_{station}}{\sum_{i=1}^{i=n} (E_G)_i} \times 100 \qquad (2.2)$$

Many of the thermal power stations compute the AP (%) as

$$AP_{station} = \left[\frac{\left(\sum_{m=1}^{m=q} (E_{UAT})_l + \sum_{j=1}^{j=p} (E_{ST})_j \right)}{\sum_{i=1}^{i=n} (E_G)_i} \right] \times 100 \qquad (2.3)$$

where E_{UAT} is the energy consumption at an individual UAT bus (measurement on 6.6 kV bus) (MWh/month) and m is the number of UAT buses in the station (m varies from 1 to q UAT).

2.3.2 Unit AP

Unit AP is the AP used at a particular unit that depends directly on plant load. Unit AP is the combination of in-house AP and common AP proportioned for that particular unit.

Computing the AP for individual units is rather difficult because it is difficult to proportion the common AP (like power used by CHP, AHP, WTP, air compressors, air-conditioning, lighting, etc.) on all the individual units. At some power plants, the total station AP is divided equally on all units. Therefore, the AP (MW) used by an individual unit is computed as

$$P_{unit} = \left(\sum_{r=1}^{r=2} (P_{UAT})_l + \frac{\sum_{j=1}^{j=p} (P_{ST})_j}{n} \right) \qquad (2.4)$$

where P_{UAT} is the average power measured at a 6.6 kV UAT bus (MW), r is the number of UAT buses, P_{ST} is the average power measured at a 6.6 kV ST bus (MW), p is the number of ST buses, and n is the number of units in that station or stage.

The total AP varies between 8.74% of gross generation at MCR condition and 11.26% of gross generation at 70% PLF (see Figure 2.12). As the PLF increases, the AP decreases. Operating the plant at 70% PLF will increase the AP by 2.52% of gross generation [32].

Figure 2.13 shows the pie diagram of AP used by different equipment at MCR condition. The AP can be broadly classified into in-house AP and common (outlying) AP.

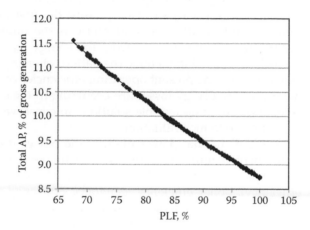

FIGURE 2.12
Variation of total AP with PLF.

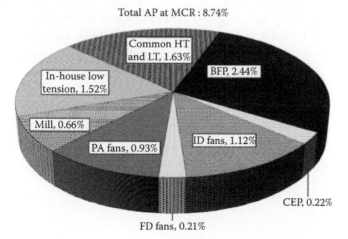

FIGURE 2.13
AP used by different equipment.

2.4 Conclusions

Electrical energy is the most popular form of energy being used, and the CO_2 emission from electrical power generation in India is about 13,393 million t/year (41.2% of total CO_2 emission), out of which 86.9% of CO_2 emission is due to power generation through fossil fuel–based technology.

Energy conservation is the technique to save and optimize energy consumption through enhancing the energy efficiency of equipment and processes. Energy conservation consists of two phases: the first phase is the conduct of an energy audit and the second phase is the implementation of energy conservation measures derived from the energy audit work. An energy audit is a technique used to identify the pattern of energy consumption and get information on the present operating efficiencies of equipment, identifying how and where energy losses are occurring in the system, analyzing the system, and providing feasible solutions to conserve the energy with a detailed techno-economic evaluation.

The AP in a coal-fired thermal power plant is the power used to drive the auxiliary equipment required to start and run the power plants. The energy performance parameters, like AP, SFC, specific oil consumption, heat rate (turbine efficiency), boiler efficiency, generator efficiency, and so on, depend on the PLF. Again the PLF depends on the fuel quality, availability of resources, forced outages, grid conditions, performance of individual equipment, and so on. The PLF of thermal power plants varies widely for different plant capacity. The AP used for a coal-fired station varies with the size of the power plant, that is, it varies between 5.2% (500 MW unit) and 12.3% (30 MW unit) at 100% MCR.

The implementation of energy conservation measures reduces the average AP of in-house AP by 1.6% of gross generation and for common AP reduces the average AP by 0.4%–0.7% of gross generation with a payback period of 1–5 years.

3

Energy Conservation of In-House Auxiliary Power Equipment in Power Plant Processes

Rajashekar P. Mandi

School of Electrical and Electronics Engineering, REVA University, Bangalore, India

Udaykumar R. Yaragatti

Department of Electrical and Electronics Engineering, National Institute of Technology Karnataka (NITK), Surathkal, India

In-house auxiliary power (AP) is used to drive the in-house equipment used for that particular unit. The power supply to this equipment is fed from unit auxiliary transformers (UATs). In-house AP is subclassified into in-house high tension (HT) and in-house low tension (LT). The total in-house AP varies between 6.5% and 8.3% of gross power generation including excitation power of 0.22%–0.30% (power used for the static excitation system of the generator) and losses in the generator transformer (GT) of 0.25%–0.35%. The total in-house AP forms about 72.5%–81.3% of the total AP of the plant.

The in-house AP (kW) is computed as

$$P_{\text{IAP}} = \sum_{i=1}^{i=n} (P_{\text{UAT}})_i \qquad (3.1)$$

where P_{UAT} is the average AP measured at the UAT bus (kW) and n is the number of UAT buses.

The total average in-house AP is about 7.11% of gross generation at maximum continuous rating (MCR) condition for 210 MW power plants. The in-house AP consists of in-house HT AP, that is, power used by major HT AP equipment and in-house LT AP.

3.1 In-House HT Equipment

The major in-house HT equipment is powered with 6.6 kV for a 210 MW plant or 11 kV for a 500 MW plant power supply and fed from the UAT bus. The average in-house HT AP forms about 5.6% of gross generation for a 210 MW plant. The loading of these HT motors directly depends on the plant load on the individual units. The major in-house HT equipment is the following:

1. Boiler feed pumps (BFPs)
2. Condensate extraction pumps (CEPs)
3. Induced draft (ID) fans
4. Forced draft (FD) fans
5. Primary air (PA) fans
6. Mills

Entries (1) to (5) above handle fluid flow, whereas (6) handles the solid, that is, coal, flow. As per the affinity law for centrifugal fluid flow elements, the fluid flow, pressure gain across fluid flow elements, speed, and power input are interrelated:

$$\frac{\overset{o}{m_1}}{\overset{o}{m_2}} = \frac{N_1}{N_2} = \left(\frac{H_1}{H_2}\right)^{1/2} = \left(\frac{P_1}{P_2}\right)^{1/3} \tag{3.2}$$

where $\overset{o}{m_1}$ is the initial fluid flow (m³/h or m³/s), $\overset{o}{m_2}$ is the final or changed fluid flow (m³/h or m³/s), N_1 is the initial speed (rpm or rps), N_2 is the final or changed speed (rpm or rps), H_1 is the initial head or pressure gain of fluid flow elements (m or MPa or kPa), H_2 is the final or changed head or pressure gain of fluid flow elements (m or MPa or kPa), P_1 is the initial power input to fluid flow elements (kW), and P_2 is the final or changed power input to fluid flow elements (kW).

From the above affinity laws, the relations of pressure gain and fluid flow with respect to power input to fluid flow elements can be rewritten as

$$\frac{P_1}{P_2} = \left(\frac{\overset{o}{m_1}}{\overset{o}{m_2}}\right)^3 = \left(\frac{H_1}{H_2}\right)^{3/2} \tag{3.3}$$

The power input to fluid flow elements directly depends on the cube of flow rate and the cube of square root of pressure gain or head. Therefore, the input power to pump/fan motor (kW) depends on the combined efficiency, pressure gain (net head, i.e., dynamic head and velocity head) across pump/

fan, and fluid flow. But all these independent parameters like pressure gain, fluid flow, and efficiency are also interrelated:

$$P_{in} \; \alpha \; \frac{PR \times \overset{o}{m}}{\eta_o} \tag{3.4}$$

where P_{in} is the power input (kW), PR is the pressure gain across fluid flow elements (kPa), $\overset{o}{m}$ is the fluid flow (m/s), and ηo is the combined efficiency of pump and motor (%).

Figure 3.1 shows the interrelation between pressure gain (head), fluid flow, efficiency of equipment, pressure drop across individual hydrodynamic resistive elements, and fluid flow leakages. The hydrodynamic resistance in fluid flow circuits influences the pressure gain across pump/fan that will vary the power input to equipment. The fluid flow leakages like passing of feed water (FW) flow through recirculation (RC) valve will alter the fluid flow handled by the equipment thereby influencing the power input to equipment. The equipment efficiencies like pump/fan efficiency, motor

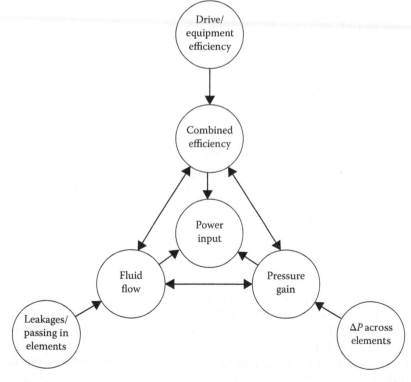

FIGURE 3.1
Interrelation between independent variable parameters of equipment.

efficiency, and drive efficiency (i.e., scoop coupling efficiency) will affect the combined efficiency, which will influence the power input to equipment. The deviation in power input, pressure gain, fluid flow, and combined efficiency of pumps/fans varies widely. The electrical power input to motor terminals is also directly related with mechanical power output along with motor and pump/fan efficiency (i.e., combined efficiency). But all these parameters will not have a similar kind of variation trend at different plant load conditions.

Table 3.1 gives the details of major in-house HT equipment for a typical 210 MW power plant and the performance results of major in-house equipment are discussed in Sections 3.1.1 through 3.1.6.

3.1.1 Boiler Feed Pumps

BFPs are the major energy-consuming equipment and are essential to increase the FW pressure in a coal-fired thermal power plant. Generally in a thermal power plant, BFPs are supplied along with booster pumps, which are mounted on the same shaft. The booster pump increases the FW pressure from 0.5 to 0.66 MPa (deaerator pressure) to intermediate pressure of about 1.2–1.4 MPa, and the BFP main pump increases the FW pressure from booster pump discharge to about 17–18 MPa (drum pressure). The average AP used by BFP will be 2.44% of total power generation and 27.9% of total AP at MCR.

BFPs are axial split multistage, horizontal, barrel type, high capacity, high speed (about 5000 rpm), and centrifugal pumps. There are three BFPs with HT induction motors of 6.6 kV and the motor rating will be either 4.0 MW or 3.5 MW in a 210 MW power plant [33]. Two pumps are working continuously and the third pump is standby. The FW flow will be regulated by scoop (fluid coupling) control and a three-element feed control valve station. Generally, two types of driving systems are used for BFPs, that is, a steam-operated turbine-driven system or motor-driven system. In a power plant with a rating of 500 MW and above, turbine-driven (steam-operated) BFPs (TDBFP) are used because the motor size will be very big, of the order of about two 10 MW motors. The starting current of these motors will be very high and influence the voltage and other power supply parameters in the network. The auxiliary steam at cold reheat line will be used to run TDBFP. This steam is already taking part in producing partial output power in the high-pressure turbine (HPT). The overall efficiency of conversion from thermal energy (coal) to hydraulic output at the BFP output is higher in the case of TDBFP compared to motor-driven BFP (MDBFP). The average conversion efficiency of coal to hydraulic power in TDBFP is about 62%, whereas in MDBFP it is 26%. But in a 210 MW power plant and lower size units, they adopt MDBFP due to lower operation and maintenance cost and also to optimize the space utilization.

Figure 3.2 shows the schematic diagram of a feed water (FW) circuit, and the main purpose of a BFP is to increase the FW pressure to meet the main

TABLE 3.1

Major In-House Auxiliary Power Equipment

S. No.	Equipment	No. of Equipment Total (Operating + Standby)	Electrical Motor Rating (Kw)	Type of Equipment	Fluid Flow (t/h)	Design Net Head (MPa)	Controls
01	BFP	3(2+1)	3500–4000	Multistage, horizontal, barrel type, high capacity, high speed, and centrifugal pumps	357.9	21.83	Hydraulic scoop
02	CEP	2(1+1)	500	Vertical, centrifugal	617.5	1.98	Throttle
03	ID fans	2(2+0)	1250–1500	Radial	670.7	3.43×10^{-3}	IGV or hydraulic scoop
04	FD fans	2(2+0)	650–750	Axial	410.3	5.10×10^{-3}	Blade pitch
05	PA fans	2(2+0)	1100–1250	Radial	262.1	11.96×10^{-3}	IGV
06	Mills	6(5+1)	340	Raymond Bowl mills XRP 863	Coal:27.8 Air:53.5	–	–

BFP, boiler feed pump; CEP, condensate extraction pump; FD, forced draft; ID, induced draft; IGV, inlet guide vane; PA, primary air.

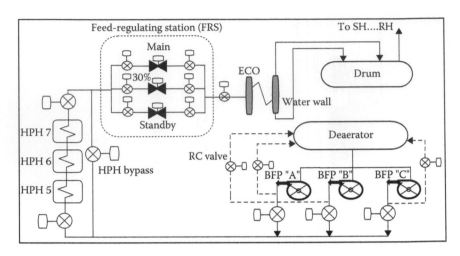

FIGURE 3.2
Schematic diagram of feed water (FW) circuit.

steam (MS) pressure (superheated steam) at the turbine inlet, that is, the HPT and intermediate-pressure turbine (IPT). While transferring the state from FW to steam, the FW has to flow through various elements like a high-pressure heater (HPH), feed-regulating station (FRS), economizer (ECO), water walls, superheaters (SHs), and reheaters (RHs), which cause the hydrodynamic resistance for FW flow. The BFP has to overcome the pressure drop across all these elements. To evaluate the performance of a BFP along with a motor, the power loss in the motor, pumps, hydrodynamic resistive elements like the HPH, FRS, ECO, water walls, SH, and RH are evaluated.

Figure 3.3 shows the AP used in different components of a BFP and FW circuit. The major power is the useful power output, that is, power available at MS, which is being fed to the turbine for converting thermal energy to mechanical energy. The major loss in the FW circuit is pump loss that forms about 23.3% of total power input (0.57% of gross generation) at MCR condition and 31.7% at 70% plant load factor (PLF). The motor loss forms about 6.4% at full load and 7% at 70% PLF. The power loss in hydraulic scoop (i.e., fluid coupling connected between pump and motor to increase the pump speed nearly to 5000 rpm at full load) is 2.8% at full load and is about 19.3% at 70% plant load. The major power loss due to hydrodynamic resistance in the boiler circuit—that is, water walls, SH and RH tubes—is about 5.7% of gross generation at MCR condition. This pressure drop across the boiler circuit depends on the scaling in the tubes due to higher silica content in FW. The power loss in HPHs is about 2.1% of gross generation, which again depends on the scaling in HPH tubes. The power loss in the FRS is about 2.4%, which depends on the pressure drop across the FRS. The power loss in ECO coils is about 0.8%, which is very low due to a lower number of tubes used compared with water wall, SH, and RH tubes.

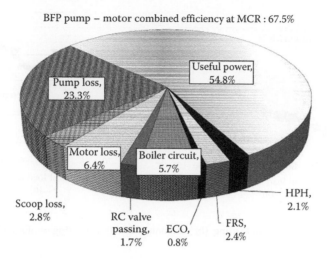

FIGURE 3.3
Auxiliary power (AP) used in boiler feed pump (BFP) and FW circuit.

As the plant load on the unit increases, the discharge pressure at the pump increases to provide the necessary steam pressure at the turbine inlet. Figure 3.4a and b shows the pressure gain across the pump and FW flow with plant load. At 100% plant load, the average measured pressure gain is 79.6% of design pump capacity (operating pressure gain margin: 20.4%). The design value of pressure gain across the BFP at 100% plant load is 82.9% (design pressure margin: 17.1%). The pumps are oversized and the operating margin is slightly on the higher side. Operation of these pumps at a nonoptimal operating point causes reduction in efficiency of pumps that increase the loss. The power loss due to operating the pump pressure gain at design condition is 0.05% of gross generation (0.5 kW/MW of plant), and the power loss due to measured average operating point at 100% PLF is 0.27% (2.7 kW/MW of plant). The lower operating pressure gain compared with design value at BFPs is mainly because of the higher margin provided for BFPs to provide higher operational reliability during single cycle operation (one set of pumps and one set of fans to raise the plant load up to 60% of plant capacity). The measured average pressure gain at 70% PLF is 77% of pump capacity and is lower than the design value of 82.8% of pump capacity. The deviation in pressure gain for operating the plant at 70% PLF in comparison with operating at MCR condition is 3.3%.

At 100% plant load, the average measured FW flow is 92.9% of pump capacity (operating flow margin: 7.1%) and is slightly more than the design value of 89% of pump capacity at 100% plant load (design flow margin: 11%). The margin provided for FW flow for the BFP is normal. The higher measured FW flow compared to the design value may be due to operating BFPs at lower pressure gain, increased specific steam consumption (SSC) of turbines, higher demineralized (DM) water makeup, use of higher auxiliary steam for tracing

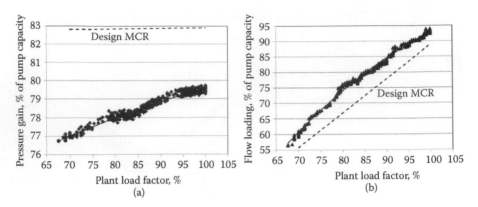

FIGURE 3.4
(a) Variation of pressure gain loading. (b) Variation of FW flow loading of BFP with plant load factor (PLF).

lines, soot blowers, and so on. The measured average FW flow at 70% PLF is 60.8% of pump capacity, which is higher than the design value of 55.8% of pump capacity. The deviation in FW flow for operating the plant at 70% PLF in comparison with operating at MCR condition is 34.6%.

As the plant load increases, the AP of BFP increases in absolute magnitude but decreases in specific value (i.e., as percentage of gross generation) (see Figure 3.5). At MCR condition, the average measured AP is 2.44% of gross generation (load factor of motor: 73.3% and margin of motor: 26.7%). The design AP at rated output capacity is 2.28% of gross generation (load factor of motor: 68.4% and margin of motor: 31.6%). The measured power input is high compared with the design value at MCR condition because the main pump's operating point is shifted due to lower efficiency of pumps, higher FW flow, and lower BFP discharge pressure. The average AP at 70% PLF is 2.96% (motor load factor: 60.8%) and is higher than the design value of 2.63% (motor load factor: 55.2%). The deviation in AP for operating the plant at 70% PLF is 21.3%, which is higher compared with other parameters.

The combined efficiency of fluid flow elements is computed as

$$\eta_o = \frac{\mathrm{PR} \times \overset{o}{m}}{P_{in}} = \eta_m \times \eta_f \tag{3.5}$$

where PR is the pressure gain (kPa), $\overset{o}{m}$ is the fluid flow (m/s), P_{in} is the power input (kW), η_m is the efficiency of the motor (%), and η_f is the efficiency of the fluid flow elements (%).

The combined efficiency (motor and pump) and specific energy consumption (SEC) of the BFP are plotted with PLF and are presented in Figure 3.6a and b, respectively. At MCR condition, the average measured combined efficiency is 67.5%, which is lower than the pump (combined) maximum efficiency of

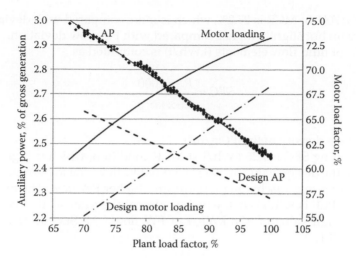

FIGURE 3.5
Variation of AP and loading of BFP motors with PLF.

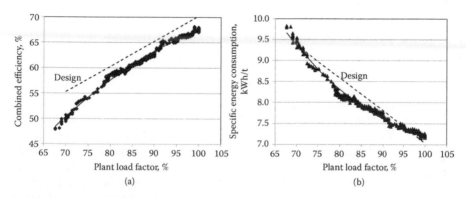

FIGURE 3.6
(a) Variation of combined efficiency. (b) Variation of specific energy consumption (SEC) of BFP with PLF.

77.6% (efficiency deviation of 10.1%). The pump efficiency at a design value at 100% plant capacity is 70.2%, which is also lower than the pump maximum efficiency point of 77.6% (efficiency deviation of 7.4%). The power loss at operating the unit at design MCR condition is 0.23% of gross generation (9.4% of BFP power), and the power loss compared with the average measured operating point is 0.32% (13.1% of BFP power). The combined efficiency is low because of lower discharge pressure, higher FW flow, higher RC flow through RC valve, problems in the pump like higher clearance between impeller and casing, pitting and erosion of the pump impeller, and so on. The average combined efficiency at 70% PLF is 50.4%, which is lower than the design value of 55.3%. The deviation in combined efficiency for operating the plant at 70% of MCR is 25.2%, which is more compared with deviation in pressure gain but slightly higher compared with FW flow deviation.

The SEC of fluid flow elements (kWh/t) is computed as

$$\text{SEC} = \frac{P_{\text{in}} \times 1000}{3.6 \times \overset{o}{m} \times \rho} \tag{3.6}$$

where ρ is the density of fluid (kg/m^3).

The operating SEC is 7.26 kWh/t at MCR condition, which is higher compared with the design value of 7.04 kWh/t of FW flow due to poor pump efficiency, lower pump discharge pressure, and higher FW flow. The measured SEC at 70% PLF is 9.36 kWh/t, which is almost the same as that compared with the design value of 9.36 kWh/t of FW flow. The SEC decreases with increase in PLF.

3.1.1.1 Energy Conservation Measures

3.1.1.1.1 Passing in RC Valves

Since the BFP is the multistage high-pressure pump, during startup of the pump, the FW flow is bypassed to the deaerator through the RC valve. During normal operation of the plant, the RC valve will be closed through an automatic signal and total FW will be used for the conversion of FW to steam. But due to very high differential pressure (DP) across the RC valve, the FW will be passing through the valve seating and the bypassed FW will return back to the deaerator. This bypassed FW will recirculate between the deaerator and BFP without taking an active part in the process. Due to passing in these valves, the FW flow at BFP suction is increased, which causes additional power input to the BFP motor. In one of the typical power plants it was measured that the FW flow (by using an ultrasonic flow meter) passing through the RC valve was in the range of 10%–15%. The additional AP of two BFP motors was 8.1% (0.2% of gross generation). The replacement of the valve seat of the RC valve for both pumps of a typical 210 MW power plant reduced the AP of the BFP by 0.30 MU/month. The investment for replacing the valve seat of the RC valve is $0.06 million and the simple payback period is 4 months, which is a very good technoeconomical feasible solution.

3.1.1.1.2 Replacement of Pump Cartridge

BFP pump efficiencies are measured in the range of 57.1%–57.4% at a typical power plant, and these BFP efficiencies are on the lower side compared with the design or predicted pump efficiency of 67.4%. The pump efficiency is low due to more clearances inside the pump, deformation of impellers, passing between interstage, and so on. In a typical power plant, the pump impeller sets (cartridge set) are replaced in both BFPs. The replacement of the BFP cartridge has enhanced pump efficiency by an average of 7%, which has reduced the power of the BFP by 0.4 MU/month for a 210 MW power plant. The investment for replacing the pump cartridge is $0.07 million and the simple payback period is 5 months. The reduction in AP of BFP is 13.1% (0.32% of gross generation), and this energy conservation measure is a very attractive solution for enhancing the energy efficiency of BFP.

3.1.1.1.3 Reduction of Pressure Drop across the FRS

The FW from the HP heater enters the FRS, where the FW pressure is regulated to maintain the drum level and drum pressure. Boiler drum level control is being practiced by two techniques, that is, DP mode (i.e., maintaining the DP across the FRS) or three-element control mode (i.e., drum level, FW flow, and MS flow). Generally, in many power plants, DP mode is being practiced where the BFP scoop maintains the drum level as per the DP set for the FRS, which is set in the range of 0.7–1.0 MPa. This leads to power loss due to DP across the FRS in the range of 0.07%–0.09% of gross generation for two BFPs in a typical 210-MW power plant. Whereas in three-element mode, the scoop of the BFP maintains the drum level as per the three-element error keeping the FRS control valve wide open. In three-element control, the pressure drop across the FRS will be less than 0.1 MPa. The average pressure drop across the FRS in many 210 MW power plants is measured in the range of 0.35–0.45 MPa. Reducing the pressure drop across the FRS from 0.35 to 0.10 MPa by operational optimization reduces the AP of the BFP by 1.7% (0.04% of gross generation).

3.1.2 Condensate Extraction Pumps

CEPs are the vertical type of centrifugal pumps. In a typical thermal power plant, two pumps are installed, where one pump is working continuously and the other is a standby. These pumps increase the condensate pressure from the condenser (i.e., vacuum 10–15 kPa) to deaerator pressure (0.5–0.7 MPa) to overcome the hydrodynamic resistances offered by gland steam condensers (GSCs) and low-pressure heaters (LPHs) (regenerative FW heaters). The average AP used by the CEP will be 0.22% of total power generation and 2.5% of total AP input at MCR. Figure 3.7 gives the schematic diagram of a condensate circuit. Figure 3.8 shows the AP used in different components of the CEP and condensate circuit. The major power is the useful power output, that is, power available at the deaerator. The major loss in the condensate circuit is pump loss that forms about 18.2% of the total power input (0.04% of gross generation) at MCR condition and 19.5% at 70% PLF. The motor loss forms about 8% at full load and 8.5% at 70% PLF. The loss in throttle valve to control the condensate pressure at the deaerator is 11.7% at full load and 22.6% at 70% PLF. Therefore, loss in the throttle valve is more at partial plant load operation. The major power loss due to hydrodynamic resistance in LPH is 6.1% at MCR condition and 6.2% at 70% PLF.

As the plant load on the unit increases, the discharge pressure at the pump increases to provide the necessary condensate/FW pressure at the deaerator.

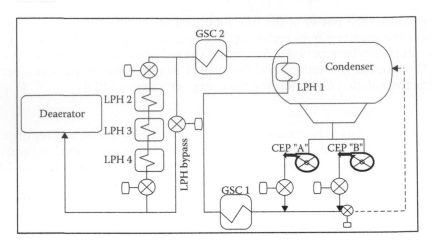

FIGURE 3.7
Schematic diagram of a condensate circuit.

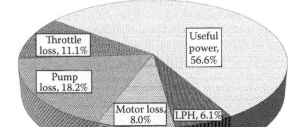

FIGURE 3.8
AP used in condensate extraction pump (CEP) and condensate circuit.

Figure 3.9a and b shows the pressure gain across the pump and condensate flow with plant load. At 100% plant capacity, the average measured pressure gain of CEP is 99.6% of pump capacity (operating pressure gain margin: 0.4%) and design pressure gain is 98.5% of pump capacity (design pressure gain margin: 1.5%). The pressure gain margin is very low and pumps are operating at their full capacity. The design and operating pressure gain of the CEP at MCR condition are optimal. The measured average pressure gain at 70% PLF is 97.2% of pump capacity and the design value is 98% of pump capacity. The deviation in pressure gain for operating the plant at 70% PLF in comparison with operating at MCR condition is 2.3%, which is very low.

In a typical power plant, the average measured condensate flow at 100% plant capacity is 80% of pump capacity (operating flow margin: 20%) and the design condensate flow is 78% (design flow margin: 22%). The operating

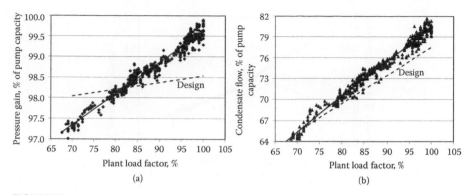

FIGURE 3.9
(a) Variation of pressure gain loading. (b) Variation of condensate flow loading of CEP with PLF.

condensate flow is high compared with the design value due to plants operating at higher SSC and higher DM water makeup. The average power loss due to higher condensate flow at MCR condition is about 0.01% of gross generation (4.5% of CEP power). The measured average condensate flow at 70% PLF is 65.3% of pump capacity and at the design value is 65.4% of pump capacity. The deviation in condensate flow for operating the plant at 70% PLF with MCR condition is 7.6%, which is higher compared with the deviation in pressure gain but is at par with deviation in power.

As the plant load increases, the AP of the CEP increases in magnitude but the specific AP as a percentage of gross generation decreases (see Figure 3.10). At MCR condition, the average measured AP is 0.22% of gross generation (load factor of motor: 85.5% and margin of motor: 14.5%). The design AP at rated output capacity is 0.21% of gross generation (load factor of motor: 80% and margin of motor: 20%). The measured power input of the CEP is high because of higher pressure drop across the LPH, lower pressure gain in the pump, higher SSC and DM water makeup flow, lower combined efficiency, and so on. The average measured AP at 70% PLF is 0.3% of gross generation (load factor of motor: 79.3% and margin of motor: 20.7%) and is higher than the design value of 0.27% of gross generation (load factor of motor: 72.6% and margin of motor: 27.4%). The deviation in AP for operating the plant at 70% PLF with MCR condition is 32.4%, which is high compared with other parameters like deviation in pressure gain and flow.

The combined efficiency (motor and pump) and SEC of the CEP are plotted with PLF, and are shown in Figure 3.11a and b, respectively. At MCR condition, the average measured combined efficiency is 62.7% and is lower compared with the pump (combined) maximum efficiency of 74.9%. The pump efficiency at the design value at 100% plant capacity is 63%, which is also lower than the pump maximum efficiency point of 74.9%. The power loss for operating the plant at design MCR condition is 0.04% of gross generation (17.9% of CEP power). The combined efficiency is low because of lower

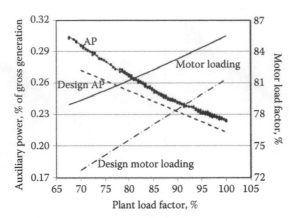

FIGURE 3.10
Variation of AP and loading of CEP motors with PLF.

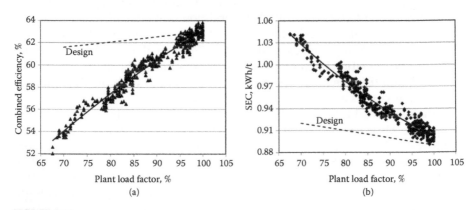

FIGURE 3.11
(a) Variation of combined efficiency. (b) Variation of SEC of CEP with PLF.

discharge pressure, higher condensate flow, higher SSC, higher DM water makeup flow, problems in the pump like higher clearance between impeller and casing, pitting and erosion of the pump impeller, and so on. The average combined efficiency at 70% PLF is 54%, which is lower than the design value of 61.6%. The deviation in combined efficiency for operating the plant at 70% PLF is 13.9%, which is slightly less compared with deviation in AP.

The operating SEC is 0.90 kWh/t at MCR condition, which is slightly higher compared with the design value of 0.89 kWh/t of condensate flow due to poor pump efficiency, lower pump discharge pressure, and higher condensate flow. The measured SEC at 70% PLF is 1.03 kWh/t, which is higher compared with the design value of 0.92 kWh/t of condensate flow. The SEC decreases with increase in PLF.

3.1.2.1 Energy Conservation Measures

3.1.2.1.1 Optimizing SSC

At many of the power plants, the SSC of plants is higher, at about 3.2–3.3 t/MW of plant output at PLF of between 90% and 100%, due to poor turbine efficiency, higher auxiliary steam consumption, and so on (see Figure 3.12). But again, as the PLF of the plant decreases, the SSC reduces to about 3.65 t/MW at 35% PLF. At higher SSC, the condensate flow will increase to produce more steam. The higher condensate flow increases the power input to the motor. Reducing the SSC from 3.3 t/MW to near design the value of 2.95 t/MW will reduce the AP of the CEP by 12.5% (0.03% of gross generation).

3.1.2.1.2 Optimizing DM Water Makeup

At many power plants, the DM water makeup is higher, in the range of 2%–3% (design value: 1%), due to the increase in continuous and intermittent blowdowns to maintain the proper silica in MS. At higher DM water makeup, the condensate flow increases. The higher condensate flow increases the power input to the motor. Reducing the DM water makeup from 2.5% to 1% will reduce the AP of the CEP by 1.8% (0.004% of gross generation).

3.1.3 ID Fans

ID fans are of the radial flow type. The main purpose of these fans is to suck the flue gas from the furnace and throw it to the atmosphere through a chimney. There are two ID fans and both are working continuously without any standby. ID fans have to essentially always maintain the furnace pressure on the negative side (draft) in the range of −49 to −98 Pa to avoid flames or hot flue gas from coming out of the furnace, which is very dangerous and hazardous for the operators. The positive pressure of furnace may also lead to explosion of the furnace. ID fans have to create suction pressure to overcome

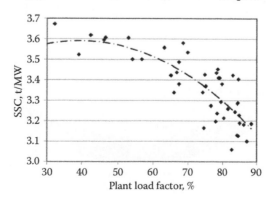

FIGURE 3.12
Variation of specific steam consumption (SSC) with PLF of typical power plant.

the hydrodynamic resistance offered by superheaters, reheaters, ECOs, air preheaters (APHs), and electrostatic precipitators (ESPs). Figure 3.13 gives the schematic diagram of a flue-gas circuit. ID fans have to handle large quantities of flue gas at a higher temperature. The flue-gas flow is controlled either by an inlet guide vane (IGV) or by a scoop (hydraulic coupling) or variable frequency drives (VFDs).

Figure 3.14 shows the AP used in different components of ID fans and the flue-gas circuit. The major loss in the flue-gas circuit is fan loss, which forms about 42.8% of total power input (0.48% of gross generation) at MCR condition and 55.5% at 70% PLF. The motor loss forms about 8.0% at full load and 8.4% at 70% PLF. The major power loss due to hydrodynamic resistance in APHs is about 27.7% of power input (0.31% of gross generation), which depends on the blocking of APH baskets. The power loss in the ECO area is about 6.1%, depending on the debris accumulated in the gooseneck area. The

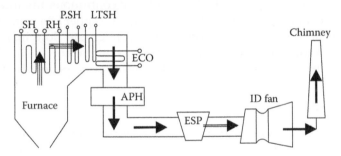

FIGURE 3.13
Schematic diagram of a flue-gas circuit.

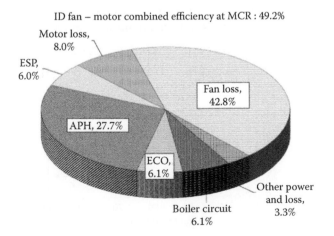

FIGURE 3.14
AP used in induced draft (ID) fans and flue-gas circuit.

power loss in an ESP is about 6.0%, depending on the ash and debris deposition in the ESP flue-gas path. The power loss in the boiler circuit, that is, the SH and RH tubes, is about 6.1% of gross generation at MCR condition.

As the plant load on the unit increases, the suction pressure (negative pressure/draft) at ID fans increases to maintain the furnace pressure at slightly negative. Figure 3.15a shows the pressure gain across ID fans with plant load. At MCR condition, the average measured pressure gain is 66.8% of fan capacity (operating pressure margin: 33.2%) and design pressure gain at 100% fan capacity is 61% of fan capacity (design pressure margin: 39%). Operation of these fans at a nonoptimal operating point causes reduction in efficiency of fans, which increases the power loss in fans. The higher pressure gain at ID fans (in comparison with the design value at MCR condition) is mainly because of the higher pressure drop across the APH, ESP, and flue-gas ducts. The power loss due to operating the fan pressure gain at a design condition is 0.04% of gross generation (3.6% of ID fan power), and the power loss due to the measured average operating point at 100% PLF is 0.06% (5.4% of ID fan power). The measured average pressure gain at 70% PLF is 61.8% of fan capacity and is higher than the design value of 36.9% of fan capacity. The deviation in pressure gain for operating the plant at 70% PLF in comparison with operating at MCR condition is 7.5%.

The flue-gas flow for both ID fans is plotted with PLF and is shown in Figure 3.15b. At MCR condition, the average measured flue-gas flow is 82.7% of fan capacity (operating flow margin: 17.3%). The design flue-gas flow at MCR condition is 75.7% of fan capacity (design flow margin: 24.3%). The higher flue-gas flow may be due to operating ID fans at lower pressure, higher excess air, higher air ingress in furnace, air leakage through the APH, air ingress through flue-gas ducts, and so on. The operating and design flow margins are on the higher side, which cause lower fan efficiency and increase the AP of ID fans. The power loss due to operating the fan at design

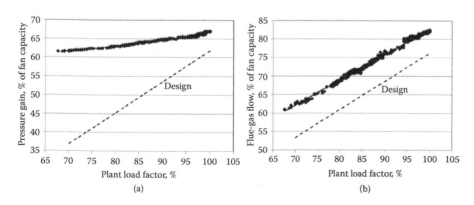

FIGURE 3.15
(a) Variation of pressure gain loading. (b) Variation of flue-gas flow loading of ID fan (IDF) with PLF.

flue-gas flow condition is 0.06% of gross generation (5.4% of ID fan power) and the power loss due to measured average operating point at 100% PLF is 0.08% (7.2% of ID fan power). The measured average flue-gas flow at 70% PLF is 62.1% of fan capacity and is higher than the design value of 53.3% of fan capacity. The deviation in flue-gas flow for operating the plant at 70% PLF in comparison with operating at MCR condition is 24.8%, which is high compared with deviation in pressure gain.

As the plant load increases, the AP of ID fans increases (see Figure 3.16). At MCR condition, the average measured AP is 1.12% of gross generation (load factor of motor: 78.2% and margin of motor: 21.8%). The design AP at rated output capacity is 0.91% of gross generation (load factor of motor: 63.7% and margin of motor: 36.3%). The measured power input is high compared with the design value at MCR condition because the operating point of fans is shifted due to lower fan efficiency, higher flue-gas flow, higher pressure drop in the flue-gas circuit, and so on. The average AP at 70% PLF is 1.32% (motor load factor: 64.7%) and is higher than the design value of 0.93% (motor load factor: 45.6%). The deviation in AP for operating the plant at 70% PLF in comparison with operating at MCR condition is 21.3%, which is higher compared with other parameters like pressure and flow.

The combined efficiency (motor and fan) and SEC of ID fans are plotted with PLF, and are shown in Figure 3.17a and b, respectively. At MCR condition, the average measured combined efficiency is 49.2%, which is lower than the fan (combined) maximum efficiency of 72.9% (efficiency deviation of 23.7%). The fan efficiency at design value at 100% plant capacity is 54.7%, which is also lower than the pump maximum efficiency point of 72.9% (efficiency deviation of 18.2%). The power loss for operating the plant at design MCR condition is 0.28% of gross generation (25% of ID fan power), and the power loss compared with the average measured operating point is 0.36% (32.2% of ID fan power). The combined efficiency is low because of lower

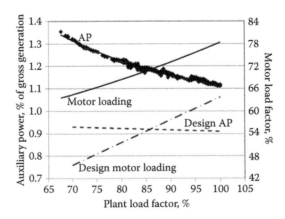

FIGURE 3.16
Variation of AP and loading of ID fan motors with PLF.

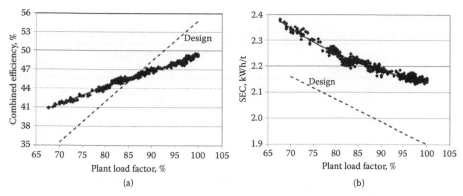

FIGURE 3.17
(a) Variation of combined efficiency. (b) Variation of SEC of IDF with PLF.

discharge pressure, higher flue-gas flow, higher excess air, higher furnace air ingress, higher air ingress in flue-gas ducts, higher flue-gas pressure drop across hydrodynamic resistive elements, poor coal quality, problems in fans like higher clearance between impeller and casing, pitting and erosion of the impeller, and so on. The average combined efficiency at 70% PLF is 41.4% and is slightly higher than the design value of 35.5%. The deviation in combined efficiency for operating the plant at 70% PLF in comparison with operating at MCR condition is 15.7%, which is lower than the deviation in flue-gas flow.

The SEC was 2.14 kWh/t at MCR condition and is higher compared with the design value of 1.90 kWh/t of flue-gas flow; this may be due to poor fan efficiency, lower fan discharge pressure, higher losses in the system and higher flue-gas flow. The measured SEC at 70% PLF is 2.36 t/h, which is higher compared with the design value of 2.16 kWh/t of flue-gas flow. The SEC decreases with increase in PLF.

3.1.3.1 Energy Conservation Measures

3.1.3.1.1 Use of Beneficiated Coal

The beneficiating of coal is nothing but removing the impurities and ash from the coal by either wet or dry washing of coal. This washing of coal reduces the raw coal size and the average ash content of coal from 52% to 31%. This also increases the calorific value of coal from 3000 to 3550 kcal/kg (Table 3.2). For the same power output, the coal flow is reduced by 8.3%, flue-gas flow is reduced by 9.4%, and ID fan power input is reduced by 8.8% (0.1% of gross generation).

3.1.3.1.2 Use of Coal Blending

Due to the deterioration of Indian coal quality over a period of time, many power plants have imported coal from different countries that has lower ash content and higher calorific value. Indian power boilers are designed with

TABLE 3.2

Performance Results of ID Fans with Beneficiated Coal for a Typical 210 MW Plant

S. No.	Particulars	Unit	Nonwashed Raw Coal	Washed Coal
01	Moisture	%	16.0	18.0
02	Ash content	%	52.0	31.0
03	Calorific value	kcal/kg	3000	3550
04	Total coal flow	t/h	128.2	117.6
05	Flue-gas flow	t/h	975.9	884.5
06	Total ID fan power	kW	2384.7	2175.8
07	Reduction in auxiliary power	kW (% of PL)	208.9 (0.10%)	

PL, plant load.

a calorific value in the range of 3500–4400 kcal/kg, but the calorific value of imported coal is in the range of 5700–6200 kcal/kg, which is higher by about 35%–40%. This higher heating value of imported coal is not suitable for direct firing in Indian boilers; also, imported coals have a low ash fusion temperature in the range of 1100–1200°C. The low ash fusion temperature of imported coal creates a clinker formation in the boiler, that is, whenever the flame temperature inside the furnace exceeds the ash fusion temperature of coal, the ash will melt and deposit on the surface of water walls, forming the clinker, which will reduce the heat transfer coefficient. Therefore, in order to utilize the imported coal, the imported coal should be used along with Indian coal to control high heat in the furnace and also overcome clinker formation in the furnace. In view of overcoming the above problems, the imported coal is fired along with Indian coal in proportion; this process is called coal blending.

As per the optimization study carried out for blending of coal, the ratio of imported coal to Indian coal can be maintained in the ratio 20% (maximum) of imported coal: 80% of Indian coal to overcome clinker formation in the furnace. However, the coal blending ratio mainly depends on the ash fusion temperature and calorific value of imported coal. The coal blending can be done in three different ways [34]:

1. *Stockpile blending*: Mixing of coal in the coal stockyard through a dozer.
2. *Belt conveyor blending*: Two separate conveyors carry Indian coal and imported coal separately and mix in a proportion by adjusting the speed of conveyors just before entering the coal bunkers [35].
3. *Tier blending*: There will be six elevations (also called six tiers) for six mills and six bunkers for a typical 210 MW power plant. Five elevations will be in service during normal operation. In tier blending, one of the bunkers, generally one of the middle bunkers, that

is, either B or C, will be fed with imported coal and the rest are fed with Indian coal. The tier blending method is generally practiced in Indian boilers because it is convenient.

The use of imported blended coal in a typical power plant has improved the net average calorific value from 2800 to 3940 kcal/kg, the coal flow is reduced by 10.3%, flue-gas flow is reduced by 7.9%, and the AP of ID fans is reduced by 7.5% (0.09% of gross generation) (see Table 3.3).

3.1.3.1.3 Control of Excess Air

While designing the power boiler or furnace of coal-fired power stations for appropriate combustion, the boilers are allowed 20% excess air in the furnace to convert all the carbon available in coal into carbon dioxide (CO_2) and only some carbon monoxide (CO), that is, less than 100 ppm is allowed. The excess air (%) in the boiler is estimated/computed by measuring the oxygen content in flue gas at the APH inlet and is computed by

$$\text{Excess air} = \frac{O_{2-APHin}}{\left(21 - O_{2-APHin}\right)} \times 100 \tag{3.7}$$

The optimum oxygen (O_2) content at the APH inlet will be 3.5%. In many power plants, the power plant operators maintain a higher FD fan blade pitch position to provide more air for combustion and also to provide more PA to lift the coal from mills to burners. This causes higher excess air in the furnace. In some other plants, the oxygen measuring port will not provide the average representative oxygen measurement possibly due to improper placement of the zirconium oxygen probe in the furnace before the APH. This misleads operators and may cause them to run the plant with higher excess air. The higher excess air increases the dry flue-gas loss in the boiler as well as increases the AP of ID fans, FD fans, and PA fans. In a typical power plant, the oxygen content at the APH inlet was measured in the range of 4.12%–4.30% (excess air: 24.4%–25.7%), which is higher compared with the design value of 3.5% (excess air: 20%) (see Table 3.4). The higher oxygen content at the APH inlet leads to higher excess air, which will increase the loading

TABLE 3.3

Performance Results of ID Fans with Blended Coal at a Typical 210 MW Plant

S. No.	Particulars	Unit	Raw Coal	Blended Coal
01	Average calorific value	kcal/kg	3000	3940
02	Total coal flow	t/h	128.2	115.0
03	Flue-gas flow	t/h	975.9	902.8
04	Total ID fan power	kW	2384.7	2206.0
05	Reduction in total power	kW (% of PL)	178.7 (0.09%)	

TABLE 3.4

Performance Results of ID Fans (IDFs) at a Typical 210 MW Power Plant

S. No.	Particulars	Unit	IDF A	IDF B
01	Plant load (PLF)	MW (%)	205.5 (97.86)	
02	Motor rating	kW	1500	
03	Pressure gain	kPa	2.832	2.954
04	Flue-gas flow	m³/s	192.78	199.67
05	Oxygen at APH inlet	%	4.12	4.30
06	Oxygen at APH outlet	%	7.32	7.31
07	Oxygen at ID outlet	%	8.89	9.12
08	Air leakage in APH	m³/s	22.41	21.12
09	Air leakage in ducts	m³/s	15.52	18.27
10	Pressure drop across APH	kPa	1.38	1.19
11	Pressure drop across ESP	kPa	0.88	0.95
12	Electrical power input	kW	1267.6	1290.1
13	Load factor of motor	%	94.76	94.96
14	Combined efficiency (design fan full load efficiency: 73.05%)	%	43.07	45.72
15	Specific energy consumption	kWh/t	8.12	7.98
16	Auxiliary power	%	1.24	
17	Power saving by reducing the O_2 at APH inlet to 3.5%	kW	19.3	26.7
18	Power saving by reducing air leakage in APH	kW	63.5	62.4
19	Power saving by reducing air ingress in duct	kW	44.0	54.0
20	Power saving by reducing pressure drop across APH	kW	75.6	47.2
21	Power saving by reducing pressure drop across ESP	kW	84.8	101.8
22	Net power saving	kW	287.2	292.1
23	New auxiliary power	%	0.96	

APH, air preheater; ESP, electrostatic precipitator, O_2, oxygen; PLF, plant load factor.

of fans as well as cause higher dry flue-gas loss in boiler. Maintaining the excess air to about 20% by adjusting the O_2 content at the APH inlet to 3.5% reduces the AP of ID fans by 2% (0.02% of gross generation).

3.1.3.1.4 Control of Air Leakage in APH

Generally in modernized thermal power plants, rotary regenerative type APHs are used to recover the heat from the outgoing flue gas. The APH will be tri-sectored, where one sector is for flue gas (which will be at negative pressure), the second sector is for secondary air (SA) with positive pressure, and the third sector is for PA, which will have high pressure [36]. Both air flows will be in the counter direction to flue-gas flow in the APH. The opening for the flue-gas path in the APH is about 180° (50% of total APH volume), for SA flow it is about 130° (about 36% of total APH volume), and for PA flow it is about 50° (about 14% of total APH volume). As a part of renovation and modernization (R&M) work in a typical 210-MW power plant to reduce the pressure drop across the APH on the PA side, the opening of the PA duct is increased from 50° to 72°, whereas the opening for PA flow is about 20% instead of 14% and SA is reduced to 30% instead of 36% [37]. Generally, manufacturers provide the margin for air leakage through the APH to about 5%–7%, but due to erosion of seals, air leakage through the APH is more. In order to reduce the air leakage through the APH, presently double sealing (radial and axial seals) are used.

In a typical power plant, the oxygen content at the APH inlet was measured in the range of 4.12%–4.30% (excess air: 24.4%–25.7%) and the oxygen content at the APH outlet was measured in the range of 7.31%–7.32% (excess air: 53.4%–53.5%). The rise in oxygen content in flue gas from the APH inlet to the APH outlet was in the range of 3.02%–3.19%, which is higher compared with the optimal value of 1% (air leakage allowed: 7.3%). The higher air leakage through the APH will increase the loading of fans. Reducing the air leakage through the APH by periodic replacement of the APH seals by double seals (radial and axial seals) reduces the AP of ID fans by 5.4% (0.06% of gross generation).

3.1.3.1.5 *Air Leakage through Flue-Gas Ducts*

The flue-gas pressure is negative throughout the total flue-gas duct from furnace to ID fan inlet because the flue gas is sucked by the ID fans. Because of negative pressure in the duct, there is always a tendency of atmospheric air ingress into the duct through small holes or openings. In many power plants it is observed that manholes are kept open and measurement ports provided in the flue-gas ducts are kept open. The air ingress into ducts increases the loading of ID fans. In a typical power plant, the oxygen content at the APH inlet is measured in the range of 4.12%–4.30% (excess air: 24.4%–25.7%), at the APH outlet it was in the range of 7.31%–7.32% (excess air: 53.4%–53.5%), and at the ID fan outlet it was in the range of 8.89%–9.12% (excess air: 73.4%–76.8%). The rise in oxygen content in the flue-gas duct from the APH outlet to ID fan outlet is in the range of 1.57%–1.81%, which is higher compared with the optimal value of 0.8% (optimum air ingress in the flue-gas duct: 6.5%). Reducing the air ingress in flue-gas duct by maintaining the rise in oxygen content from an average value of 1.7–0.8% will reduce the AP of ID fans by 4.2% (0.05% of gross generation).

3.1.3.1.6 *Flue-Gas Pressure Drop across the APH*

In the APH, the heat in the flue gas is transferred from flue gas to air through the baskets. The baskets are made up of a steel metallic honeycomb-like structure, which is used to transfer the heat. Generally, soot blowers are installed just above the APH and near the ECO to clean the heating surface to enhance the heat transfer coefficient in the ECO. While operating the soot blowers, some parts of the soot-blowed steam are converted into water particles, which will mix with the fly ash present in flue gas and form a cementing effect in air baskets in the APH, which blocks the APH. This will create hydrodynamic resistance in the flue-gas and air circuits. These pressure drops increase the AP of air fans and ID fans. In a typical power plant, the pressure drop across the APH on the flue-gas side was measured in the range of 1.19–1.38 kPa, which is higher than the design value of 1.02 kPa. Reducing the flue-gas pressure drop in the APH from an average value of 1.25–1.02 kPa reduces the AP of ID fans by 5.2% (0.06% of gross generation).

3.1.3.1.7 *Pressure Drop in Flue-Gas Circuit across the ESP*

The total ash in the power boiler is generally split in the ratio of 80:20, that is, 80% of total ash is converted to fly ash, which is very small and is collected in the ESP. About 20% of the total ash (in the form of clinker) is collected at the bottom of the furnace, which is considered bottom ash. The dust-laden gas passes through an ESP that collects most of the fly ash. Cleaned gas then passes out of the precipitator and is thrown out to the atmosphere by ID fans through the chimney. Precipitators typically collect 99.9% or more of the dust from the gas stream. ESPs charge the dust particles electrostatically in the gas stream. The charged particles are then attracted to and deposited on collecting plates. When sufficient dust has accumulated on the collecting plates, the collectors are vibrated (rapped) to dislodge the dust, causing it to fall with the force of gravity on hoppers below by using the rapping mechanism. The dust is then collected at the bottom of hoppers and mixed with water for a wet ash disposal system or collected dry in a silo system. When the fluegas passes through the ESP, the flue-gas pressure drops due to hydrodynamic resistance offered by ESP components. In a typical power plant, the flue-gas pressure drop across the ESP is measured in the range of 0.88–0.95 kPa, which is higher than the design value of 0.44 kPa. If the flue-gas pressure drop across the ESP is reduced from an average value of 0.90–0.44 kPa, the AP of ID fans is reduced by 8% (0.09% of gross generation).

3.1.3.1.8 *Variable Frequency Drives*

Generally for fluid flow elements like pumps and fans, it is common practice to use constant speed induction motors with IGV or an outlet damper or hydraulic coupling to control the fluid flow. In the case of IGV and outlet damper control, the system resistance will be higher, so more energy is wasted at partial loading. In the case of hydraulic coupling, the efficiency of the hydraulic coupling is poor at a lower speed. As per the affinity rules, the power is proportional to the cube of speed, so reducing the speed of the motor at partial load will reduce the power input considerably (see Table 3.5). Since the design margin for ID fans is very high (pressure of 44%, flow of 26%, and motor of 30%), the installation of a VFD is highly economical. The installation of a VFD for ID fans reduces the AP by 0.29% of gross generation (26% of ID fan power). The anticipated investment for the replacement of ID fans by a VFD will be $0.95 million and the simple payback period will be 4 years. But these VFDs inject current harmonics in the system. The voltage

TABLE 3.5

Performance Results of ID Fans with and without a Variable Frequency Drive (VFD)

Sl. No.	Particulars	Without VFD	VFD
01	Plant load, MW	209	210
02	No. of fans/rating, kW	2/1300	2/1500
03	Suction pressure, kPa	−2.68	−2.56
04	Discharge pressure, kPa	0.078	0.074
05	Flue-gas flow, m³/s	351.52	349.9
06	Electrical power input, kW	1962.53	1353.56
07	Mechanical power output, kW	969.49	921.64
08	Operating combined efficiency, %	49.40	68.09
09	SEC, kWh/t	1.94	1.34
10	Power saving, kW	608.97	
11	Energy saving, MU/year	5.26	

SEC, specific energy consumption.

total harmonic distortion (VTHD) is in the range of 0.8%–0.9% and the individual voltage harmonics are less than 0.52%. The voltage harmonics are well within the limits prescribed by the IEEE (Institute of Electrical and Electronics Engineers) 519 standard (total harmonic distortion must be below 5.0% and individual harmonics must be less than 3.0% for a voltage level up to 69 kV). The current total harmonic distortion (ITHD) is measured in the range of 32%–33% and the individual current harmonics are in the range of 1.6%–27.1%. The current harmonic limits depend on the ratio of short circuit current of that network to peak measured current (I_{SC}/I_L). The I_{SC}/I_L ratio is 180 at a UAT bus in a typical 210 MW power plant; as per the IEEE 519 standard, the total demand distortion (TDD) must be less than 15% and individual current harmonics must be as follows: h<11: 12%, 11<h>17: 5.5%, 17<h>23: 5%, 23<h>35: 2%, and h>35: 1%. But these harmonic limits are applicable at the point of common coupling (PCC). In this power plant, the PCC will be at the secondary UAT. At the secondary UAT secondary, the maximum current TDD is measured as 3.2%, the fifth individual current harmonics is less than 2.1%, and subsequent other individual current harmonics are less than 1.1%. The current harmonics are within the limits prescribed by the IEEE 519 standard.

3.1.4 FD Fans

FD fans are of the axial flow type. In this type of fan, the air flows axially, that is, parallel to the axis of rotation. There are two FD fans and both fans are working continuously without any standby. FD fans handle atmospheric air and are low-pressure, high-volume handling fans. FD fans provide the hot SA (atomizing air) for combustion. FD fans suck the atmospheric air at ambient temperature, force the air through the APH where the air is heated, and supply the wind box (see Figure 3.18). These fans have to maintain the DP across the wind box at about 0.98 kPa (positive pressure) in order to maintain appropriate combustion. The SA flow is controlled by altering the blade pitch of FD fan impellers. Figure 3.19 shows the AP used in different components of FD fans and the SA circuit. The major power is the useful power output, that is, power available at the wind box to provide the atomizing air for combustion at the furnace, which forms about 34% of total power input (0.07% of gross generation). The major loss in the secondary circuit is fan loss, which forms about 31.3% of total power input (0.06% of gross generation) at MCR condition and 32.9% at 70% PLF. The motor loss forms about 8.5% at full load and 9.1% at 70% PLF. The major power loss due to hydrodynamic resistance in the APH is about 26.2% of power input (0.05% of gross generation), and depends on the blocking of APH baskets.

As the plant load on the unit increases, the discharge pressure at the fan increases to provide the necessary SA pressure at the wind box. Figure 3.20a and b shows the pressure gain across the FD fans and SA flow with plant load.

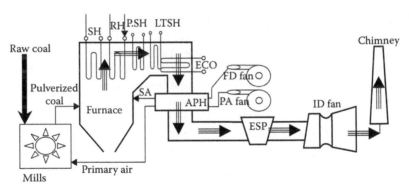

FIGURE 3.18
Schematic diagram of secondary air (SA) circuit.

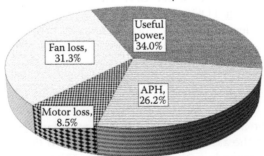

FIGURE 3.19
AP used in forced draft (FD) fans and SA circuit.

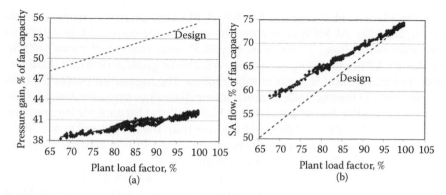

FIGURE 3.20
(a) Variation of pressure gain loading. (b) Variation of SA flow loading of FD fan (FDF) with PLF.

At MCR condition, the average measured pressure gain is 42% of fan capacity (operating pressure gain margin: 58%) and design pressure gain at 100% fan capacity is 57.2% of fan capacity (design pressure margin: 42.8%). This operation of FD fans at a nonoptimal operating point causes drastic reduction in fan efficiency that increases the power loss in fans. The power loss for operating the plant at design pressure gain at MCR condition is 0.007% of gross generation (3.3% of FD fan power), and the power loss compared at the measured operating point at MCR condition is 0.02% (9.3% of FD fan power). The lower pressure gain at FD fans (in most of the power plants) is mainly because of higher illegal furnace air ingress through the furnace openings, joints, corners, ceilings, and so on. This illegal air ingress will not directly take part in the combustion process, but increases the excess air, which increases the dry flue-gas losses in the boiler that increases the loading of ID fans. Therefore, to control the excess air in furnace, the wind box pressure is maintained below 0.98 kPa (about 100 mmWC). But lower wind box pressure leads to improper combustion and generates more CO, that is, carbon (heat) loss in the boiler. The measured average pressure gain at 70% PLF is 38.7% of fan capacity and is lower than the design value of 49.6% of fan capacity. The deviation in pressure gain for operating the plant at 70% PLF in comparison with operating at MCR condition is 7.8%.

At MCR condition, the average measured SA flow is 74% of fan capacity (operating flow margin: 26%) and is at par with the design value. The power loss for operating the plant at the design SA flow at MCR condition is 0.013% of gross generation (6% of FD fan power). The measured average SA flow at 70% PLF is 59.6% of fan capacity and is higher than the design value of 53.6% of fan capacity. The deviation in SA flow for operating the plant at 70% PLF in comparison with operating at MCR condition is 19.5% and is high compared with deviation in pressure gain.

As the plant load increases, the AP of FD fans increases in magnitude (see Figure 3.21). At MCR condition, the average measured AP is 0.215% of gross generation (load factor of motor: 32.5% and margin of motor: 67.5%). The design AP at rated output capacity is 0.25% of gross generation (load factor of motor: 35% and margin of motor: 65%). The measured power input is less because FD fans operate at much lower discharge pressure and SA flow compared with the design value due to higher illegal furnace air ingress. The average AP at 70% PLF is 0.25% and is lower than the design value of 0.266%. The deviation in AP for operating the plant at 70% PLF in comparison with operating at 100% PLF is 17.0%, which is moderate.

The combined efficiency (motor and fan) and SEC of FD fans are plotted with PLF, and are shown in Figure 3.22a and b, respectively. At MCR

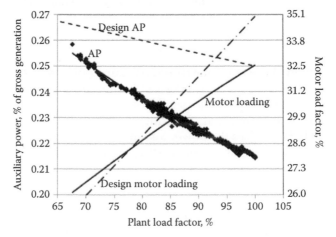

FIGURE 3.21
Variation of AP and loading of FD fan motors with PLF.

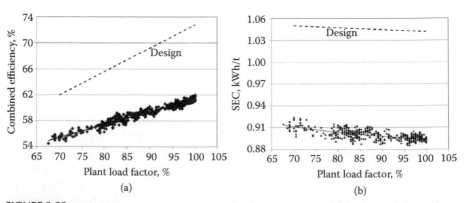

FIGURE 3.22
(a) Variation of combined efficiency. (b) Variation of SEC of FDF with PLF.

condition, the average combined efficiency is 61.3% and is lower than the fan maximum efficiency of 79.2% (efficiency deviation of 17.9%). The fan efficiency at design value at 100% plant capacity is 72.8%, which is also lower than the pump maximum efficiency point of 79.2% (efficiency deviation of 6.4%). The power loss for operating the plant at design MCR condition is 0.02% of gross generation (8.9% of FD fan power), and the power loss compared at the measured operating point at MCR condition is 0.05% of gross generation (22.3% of FD fan power). The combined efficiency is low because of problems in fans like fan oversizing, that is, shift in the operating point of fan design characteristics, change in fan blade angle, clearance between impeller and casing, pitting and erosion of the fan impeller, and so on. The average combined efficiency at 70% PLF is 55.4% and is lower than the design value of 62.0%. The deviation in combined efficiency for operating the plant at 70% PLF is 9.7%.

SEC is 0.89 kWh/t at MCR condition and is lower compared with the design value of 1.042 kWh/t of SA flow; this may be due to lower SA flow, lower fan discharge pressure, and so on. The measured SEC at 70% PLF is 0.91 kWh/t of SA flow, which is lower compared with the design value of 1.05 kWh/t of SA flow. The SEC decreases with the increase in PLF.

3.1.4.1 Energy Conservation Measures

3.1.4.1.1 Use of Beneficiated Coal

Using beneficiated coal for the same power output, the SA flow is reduced by 3.4%, which reduces the AP of FD fans by 3.9% (0.01% of gross generation) (Table 3.6).

3.1.4.1.2 Use of Coal Blending

Using imported blended coal in typical power plant, the SA flow is reduced by 1.8%, which reduces the AP of FD fans by 2% (0.004% of gross generation) (see Table 3.7).

TABLE 3.6

Performance Results of FD Fans with Beneficiated Coal for Typical 210 MW Plant

S. No.	Particulars	Unit	Nonwashed Raw Coal	Washed Coal
01	Secondary air flow	t/h	498.7	481.6
02	Total FD fan power	kW	450.3	432.9
03	Reduction in auxiliary power	kW (% of PL)	17.4 (0.01%)	

3.1.4.1.3 Control of Excess Air

Maintaining the excess air to 20% by adjusting the O_2 content at the APH inlet to 3.5% reduces the AP of FD fans by 5.1% (0.01% of gross generation) (see Table 3.8).

3.1.4.1.4 Air Leakage in APH

In a typical power plant, the oxygen content at the APH inlet is measured in the range of 4.12–4.30% and the oxygen content at the APH outlet is in

TABLE 3.7

Performance Results of FD Fans with Blended Coal for a Typical 210 MW Plant

S. No.	Particulars	Unit	Raw Coal	Blended Coal
01	Secondary air flow	t/h	498.7	489.8
02	Total FD fan power	kW	450.3	441.5
03	Reduction in auxiliary power	kW (% of PL)		8.8 (0.004 %)

TABLE 3.8

Performance Results of FD Fans at a Typical 210 MW Power Plant

S. No.	Particulars	Unit	FDF A	FDF B
01	Plant load (PLF)	MW (%)	205.5 (97.86)	
02	Motor rating	kW		750.00
03	Suction pressure	kPa	−0.118	−0.118
04	Discharge pressure	kPa	1.737	1.684
05	Pressure gain	kPa	1.855	1.802
06	Secondary air flow	m³/s	69.66	69.15
07	Oxygen at APH inlet	%	4.12	4.30
08	Oxygen at APH outlet	%	7.32	7.31
09	Air leakage in APH	m³/s	13.21	12.47
10	Pressure drop across APH	kPa	0.874	0.893
11	Electrical power input	kW	265.91	278.45
12	Load factor of motor	%	31.20	32.67
13	Combined efficiency (design fan full load combined efficiency: 79.2%)	%	48.60	44.75
14	Specific energy consumption	kWh/t	3.67	3.87
15	Auxiliary power	%		0.26
16	Power saving by controlling air leakage in APH	kW	24.5	22.5
17	Power saving by reducing the O_2 at APH inlet from 4.21% to 3.5%	kW	9.8	13.2
18	Power saving by reducing pressure drop across APH from 0.88 to 0.86 kPa	kW	3.3	4.2
19	Net power saving	kW	37.6	39.9
20	New auxiliary power	%		0.24

the range of 7.31%–7.32%. The rise in oxygen content in flue gas from APH inlet to APH outlet was in the range of 3.02%–3.19%, which is higher compared with the optimal value of 1.0% (air leakage allowed: 7.3%). Reducing the air leakage through the APH by periodic replacement of the APH seals by double seals (radial and axial seals) reduces the AP loss of FD fans by 10.4% (0.02% of gross generation).

3.1.4.1.5 SA Pressure Drop across APH

In a typical power plant, the pressure drop across APH on the SA side is measured in the range of 0.87–0.90 kPa, which is higher than the design value of 0.86 kPa. Reducing the SA pressure drop in the APH from an average value of 0.88–0.86 kPa reduces the AP of FD fans by 1.7% (0.004% of gross generation).

3.1.5 PA Fans

PA fans are of the radial flow type. There are two PA fans and both are working continuously without any standby. Generally two types of PA fans are being used, that is, some PA fans handle hot air, and are in series with FD fans. PA fans provide high-pressure hot air to lift the pulverized coal (PC) from mills to burners. PA fans have to provide the air pressure to overcome the hydrodynamic resistance offered by the APH, air ducts, and mill DP. PA fans have to provide the PA pressure at the mill inlet to an optimum value of about 6.43 kPa to overcome the pressure drop across the APH of about 0.42 kPa and average DP of about 3.64 kPa in mills at MCR condition. Figure 3.23 shows the schematic diagram of a PA circuit.

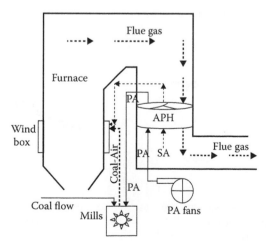

FIGURE 3.23
Schematic diagram of primary air (PA) circuit.

Figure 3.24 shows the AP used in different components of the PA fans and PA circuit. The major power is the useful power output, that is, the power to lift the coal from mills to burners to aid combustion in the furnace, which forms about 32.5% of the total power input (0.3% of gross generation). The major loss in a PA circuit is fan loss, which forms about 56.6% of total power input (0.53% of gross generation) at MCR condition and 73.7% at 70% PLF. The motor loss forms about 6.7% at full load and 7.2% at 70% PLF. The major power loss due to hydrodynamic resistance in the APH is about 4.2% of power input (0.04% of gross generation), which depends on the blocking of APH baskets.

As the plant load on the unit increases, the discharge pressure at the fan increases to provide the necessary PA pressure at the mill inlet. Figure 3.25a and b shows the pressure gain across PA fans and PA flow with plant load. At MCR condition, the average measured pressure gain is 69% of fan capacity (operating pressure gain margin: 31%). At 100% plant load, design pressure

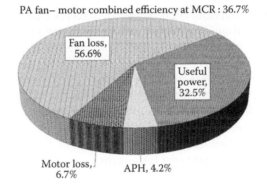

PA fan– motor combined efficiency at MCR : 36.7%

FIGURE 3.24
AP used in PA fans and PA circuit.

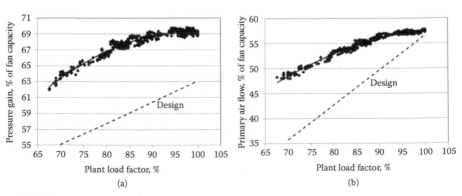

FIGURE 3.25
(a) Variation of combined efficiency. (b) Variation of SEC of PA fan (PAF) with PLF.

gain is 63% of fan capacity (design pressure gain margin: 37%). The operation of PA fans at a nonoptimal operating point causes drastic reduction in the efficiency of fans that increases the power loss. The power loss difference between the operating point with design pressure gain at MCR condition is 0.02% of gross generation (2.1% of PA fan power) and the power loss compared at the measured operating point at MCR condition is 0.05% (9.6% of PA fan power). The higher pressure gain at the PA fans is mainly because of fear of use of poor coal quality. The measured average pressure gain at 70% PLF is 63% of fan capacity and is higher than the design value of 55% of fan capacity. The deviation in pressure gain for operating the plant at 70% PLF in comparison with operating at MCR condition is 8.7%.

At MCR condition, the average measured PA flow is 57.6% of fan capacity (operating pressure gain margin: 42.4%). At 100% plant capacity, the design pressure gain is 56.5% of fan capacity (design pressure gain margin: 43.5%). The measured average PA flow at 70% PLF is 48.5% of fan capacity and is higher than the design value of 35.8% of fan capacity. The deviation in PA flow for operating the plant at 70% of MCR is 15.8% and is high compared with pressure gain.

As the plant load increases, the AP of PA fans increases in magnitude (see Figure 3.26). At MCR condition, the average measured AP is 0.93% of gross generation (load factor of motor: 78.5% and margin of motor: 21.5%). The design AP at rated output capacity is 0.49% of gross generation (load factor of motor: 40.8% and margin of motor: 59.2%). The measured power input is high because the operating point of PA fans is shifted due to lower efficiency of fans, poor coal quality, higher PA flow, and higher PA fan discharge pressure. The average AP at 70% PLF is 1.2% of gross generation (load factor of motor: 70.5%) and is higher than the design value of 0.5% of gross generation (load factor of motor: 29.1%). The deviation in AP for operating the plant at 70% PLF in comparison with operating at MCR condition is 28.2%.

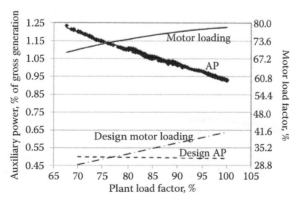

FIGURE 3.26
Variation of AP and loading of PA fan motors with PLF.

The combined efficiency (motor and fan) and SEC of PA fans are plotted with PLF, and are shown in Figure 3.27a and b, respectively. At MCR condition, the average combined efficiency is 36.7% and is lower than the fan maximum efficiency of 72.9% (efficiency deviation of 36.2%). The fan efficiency at design value at 100% plant capacity is 59.5%, which is also lower than the fan maximum efficiency point of 72.9% (efficiency deviation of 13.4%). The power loss for operating the plant at design MCR condition is 0.18% of gross generation (19.3% of PA fan power) and the power loss compared at the measured operating point at MCR condition is 0.27% of gross generation (28.9% of PA fan power). The combined efficiency is low because of problems in fans like fan oversizing, that is, a shift in the operating point of fan design characteristics, change in fan blade angle, clearance between the impeller and casing, pitting and erosion of the fan impeller, and so on. The average combined efficiency at 70% PLF is 31.3% and is lower than the design value of 48.3%. The deviation in combined efficiency for operating the plant at 70% PLF in comparison with operating at MCR condition is 14.9%.

The SEC is the ratio of power input (kW) to PA flow (t/h) handled by PA fans. At MCR condition, the average measured SEC is 6.5 kWh/t and is higher than the design value of 3.7 kWh/t at MCR condition due to use of poor quality coal. SEC at 70% PLF is 6.9 kWh/t and is higher than the design value of 4.0 kWh/t. Operating the plant at 70% PLF in comparison with operating at MCR condition, the SEC increases by about 0.43 kWh/t of PA flow.

3.1.5.1 Energy Conservation Measures

3.1.5.1.1 Use of Beneficiated Coal

Using beneficiated coal for the same power output, the PA flow is reduced by 20%, which reduces the AP of PA fans by 19.5% (0.18% of gross generation) (see Table 3.9).

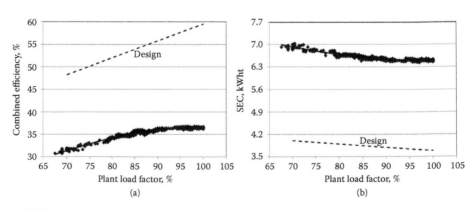

FIGURE 3.27
(a) Variation of combined efficiency. (b) Variation of SEC of PAF with PLF.

3.1.5.1.2 Use of Coal Blending

Using imported blended coal in a typical power plant, the PA flow is reduced by 16.4%, which reduces the AP of PA fans by 19% (0.18% of gross generation) (see Table 3.10).

3.1.5.1.3 Control of Excess Air

Maintaining the excess air to 20% by adjusting the O_2 content at the APH inlet to 3.5% reduces the AP of PA fans by 1.9% (0.02% of gross generation) (see Table 3.11).

3.1.5.1.4 Air Leakage in the APH

In a typical power plant, the oxygen content at the APH inlet is measured in the range of 4.12–4.30% and the oxygen content at the APH outlet is in the range of 7.31%–7.32%. The rise in oxygen content in flue gas from the APH inlet to the APH outlet is in the range of 3.02%–3.19%, which is higher compared with the optimal value of 1.0% (air leakage allowed: 7.3%). Reducing the air leakage through APH by periodic replacement of APH seals by double seals (radial and axial seals) reduces the AP loss of PA fans by 2.7% (0.03% of gross generation).

TABLE 3.9

Performance Results of PA Fans with Beneficiated Coal for Typical 210 MW Plant

S. No.	Particulars	Unit	Nonwashed Raw Coal	Washed Coal
01	PA flow	t/h	285.9	228.5
02	Total PA fan power	kW	1963.5	1579.9
03	Reduction in auxiliary power	kW (% of PL)	383.6 (0.18%)	

TABLE 3.10

Performance Results of PA Fans with Blended Coal at a Typical 210 MW Plant

S. No.	Particulars	Unit	Raw Coal	Blended Coal
01	PA flow	t/h	285.9	238.9
02	Total PA fan power	kW	1963.5	1589.9
03	Reduction in auxiliary power	kW (% of PL)	373.6 (0.18 %)	

TABLE 3.11

Performance Results of PA Fans (PAFs) at a Typical 210 MW Power Plant

S. No.	Particulars	Unit	PAF A	PAF B
01	Plant load (PLF)	MW (%)	205.5 (97.86)	
02	Motor rating	kW	1250	
03	Suction pressure	kPa	−0.069	−0.069
04	Discharge pressure	kPa	9.402	8.968
05	Pressure gain	kPa	9.471	9.037
06	PA flow	m³/s	48.53	47.98
07	Oxygen at APH inlet	%	4.12	4.30
08	Oxygen at APH outlet	%	7.32	7.31
09	Air leakage in APH	m³/s	9.20	8.65
10	Pressure drop across APH	kPa	2.46	2.19
11	Electrical power input	kW	1257.59	1260.10
12	Combined efficiency (design fan full load combined efficiency: 72.86%)	%	36.55	34.41
13	Specific energy consumption	kWh/t	6.92	7.01
14	Auxiliary power	%	1.22	
15	Power saving reducing ΔP across APH	kW	102.5	88.3
16	Power saving by reducing air leakage in APH	kW	31.2	22.3
17	Power saving by reducing the O_2 at APH inlet to 3.5%	kW	16.9	20.6
18	Power saving by reducing fan discharge pressure to 8.0 kPa	kW	146.2	123.2
19	Net power saving	kW	296.8	254.4
20	New auxiliary power	%	0.96	

3.1.5.1.5 PA Pressure Drop across the APH

In a typical power plant, the pressure drop across the APH on the PA side is measured in the range of 2.19–2.46 kPa, which is higher than the design value of 0.42 kPa. Reducing the PA pressure drop in the APH from an average value of 2.33–0.42 kPa reduces the AP of PA fans by 9.7% (0.09% of gross generation).

3.1.5.1.6 Optimizing PA Fan Discharge Pressure

The main purpose of PA fans is to deliver the primary air to overcome the pressure drop across the APH (i.e., about 0.42 kPa) and the mill differential pressure (i.e., about 3.64 kPa), and to carry the pulverized coal from the mills to the burners at the wind box. The design PA pressure at the mill inlet would be 6.43 kPa. The optimum value for PA fan discharge would be about 8.0 kPa for a typical 210 MW plant. In many power plants, the PA fan discharge pressure is maintained on the higher side in the range of 8.8–9.4 kPa. The higher discharge pressure increases the AP of PA fans. The PA fan discharge pressure is measured in the range of 8.97–9.40 kPa. If the PA fan discharge pressure is reduced to 8.0 kPa, the AP of PA fans will be reduced by 13.7% (0.13% of gross generation).

3.1.6 Coal Mills

All the coal quality parameters have a great impact on the performance of fans and mills. Over five to six decades, coal quality has deteriorated (i.e., reduction in calorific value and increase in ash content of coal) because of the depletion of better quality coal reserves. Now coal mining is done with surface mining. The main reason for increased ash content is increased opencast mining and production of coal from inherently inferior grades of coal. Many times noncoal (foreign) materials such as shale, stones, and occasionally even iron pieces (such as shovel teeth) have been found in a run of mine coal. Most Indian coal power plants burn coal without any prior cleaning. Transporting large amounts of ash-forming minerals wastes energy and creates shortages of rail cars and port facilities. Burning of low-quality, high-ash coals also creates problems for power stations, including erosion, difficulty in pulverization, poor emissivity and flame temperature, low radiative heat transfer, excessive amount of fly ash containing large amounts of unburned carbons, and so on.

For power boilers, the common coal used in Indian power plants are bituminous and subbituminous coal. The gradation of Indian coal is based on its calorific value measured at coal mines (see Table 3.12). Generally in Indian power boilers, grades D, E, and F are being used for power generation.

It is a common experience in most thermal power plants that the coal received deviates from design values in terms of calorific value as well as ash content. The calorific value of coal from the United States and China is almost twice that of Indian coal [38]. The specific fuel consumption for Indian coal is about 0.7–0.9 kg/kWh as compared with 0.4 kg/kWh for U.S. and China coal. The power used by the coal handling plant (CHP) and milling system is almost double. However, Indian boilers are designed with calorific value between 4500 and 4800 kcal/kg. The use of higher calorific value coal in Indian boilers will mismatch the heating surface, especially for water walls. The ash content of Indian coal is almost two to three times than that of imported coal.

Figure 3.28a shows the design fuel analysis data for a typical power plant, and Figure 3.28b shows the variation of monthly average coal quality parameters (proximate analysis data and calorific value) for a typical power plant

TABLE 3.12

Grading of Indian Coal

Grade	Calorific Value (kcal/kg)
A	Exceeding 6200
B	5600–6200
C	4940–5600
D	4200–4940
E	3360–4200
F	2400–3360
G	1300–2400

for a period of 3 years. The ash content in coal is increased from 39.64% to 52.44% (design value: 29%) for the period of 3 years, whereas the fixed carbon is decreased from 27.78% to 22.17% (design value: 35%), and volatile matter is also decreased from 20.76% to 15.35 % (design value: 26%). The calorific value of coal is also reduced from 4200 to 3020 kcal/kg (design value: 4400 kcal/kg) over the 3 years. This reduced coal quality has increased the AP of equipment.

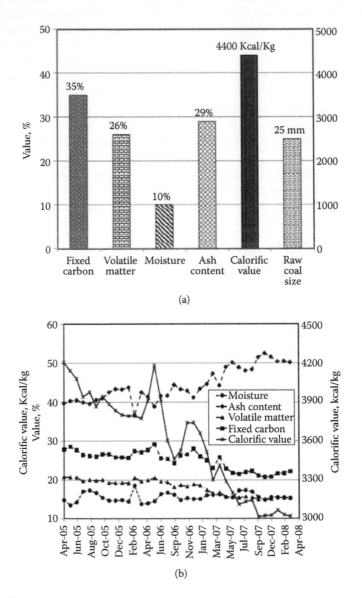

FIGURE 3.28

(a) Proximate analysis of design coal. (b) Variation of coal quality parameters at typical power.

Figure 3.29 gives the variation of AP with calorific value of coal. The calorific value of coal is reduced from 3459 to 2847 kcal/kg of coal, and AP is increased from an average value of 8.56%–10.80%. The heat rate and AP are greatly influenced by the calorific value of coal. As the calorific value of coal reduces, the AP of the plant increases.

In a coal circuit, the coal is received at the CHP through a coal wagon from the mines. The coal is crushed in crushers and the coal size is reduced to about 25 mm. In some power plants, two stages of crushing are used, that is, a primary crusher to reduce the coal size from big lumps to about 125 mm and a secondary crusher to reduce the coal size from 125 to 25 mm into raw coal. The primary and secondary crushers are of hammer or impactor type mills. The raw coal is transported to the mills through conveyor belts and the raw coal feeder. Generally, raw coal feeders are either gravimetric or volumetric. The coal flow is measured by calibrating the speed of the raw coal feeder (volumetric feeder) or load cells installed in gravimetric feeders. Mills convert raw coal into PC of the size where about 70% pass through a 75 µm size mesh. This PC is lifted by the PA to the burners for combustion. There are three types of coal mills, that is, Raymond bowl mills, large ball E-type mills, and tube and ball mills. Generally in most power plants, Raymond bowl mills are used to minimize the AP. In a typical 210 MW power plant, six bowl mills are installed. As per the design condition of coal, four mills have to work continuously, one mill will be hot standby, and the other mill will be cold standby. But due to the use of inferior coal quality in most of the power plants, five mills are working continuously and the sixth mill is a standby.

The power input to mill motors depends on the coal flow. In many power plants the coal flow is calibrated with the raw coal (RC) feeder speed, and a few plants have gravimetric feeders where the coal flow can be measured

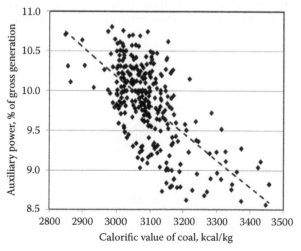

FIGURE 3.29
Variation of AP with calorific value of coal.

using load cells installed in the feeder. The gravimetric feeder gives a better coal measurement compared with the RC feeder [39]. The specific coal flow for mills is plotted with PLF in Figure 3.30. The measured average specific coal flow at MCR condition is 0.67 t/MW and is higher than the design value of 0.64 t/MW. The higher specific coal consumption is mainly due to the use of inferior quality coal. The specific coal flow when operating the plant at 70% PLF is 0.74 t/MW and is higher than the design value of 0.66 t/MW. The deviation in specific coal flow for operating the plant at 70% PLF in comparison with operating at MCR condition is 10.4%.

In order to evaluate the AP used by mills, it is computed as the ratio of mill power input to plant load. The variation of AP and SEC for all mills are plotted with variation in plant load, and are shown in Figure 3.31a and b, respectively. It can be seen from the figure that as the plant load increases, the AP of mills decreases. The AP at MCR condition is 0.66% of gross generation (motor load factor: 82.1% and margin: 17.9%), whereas at 70% PLF, the AP is 0.86% of gross generation (motor load factor: 74.1% and margin: 25.9%). The deviation in AP for operating the plant at 70% PLF in comparison with operating at MCR condition is 30.3%.

The SEC is one of the performance parameters for mills. The SEC varies between 9.7 kWh/t of coal flow at MCR condition and 11.5 kWh/t of coal flow at 70% PLF. The difference in SEC for operating the plant at 70% PLF in comparison with operating at MCR condition is 1.8 kWh/t. The SEC decreases with increase in PLF.

3.1.6.1 Energy Conservation Measures

3.1.6.1.1 Use of Higher Calorific Value Coal

Nowadays, many Indian power plants are using lower calorific value coal, which increases the AP. The use of improved coal quality, that is, increase in

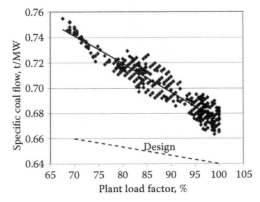

FIGURE 3.30
Variation of coal flow with PLF.

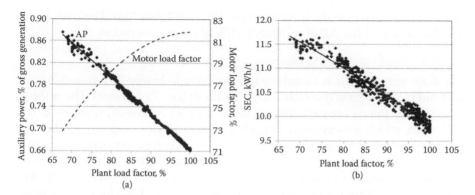

FIGURE 3.31
(a) Variation of AP and mill motor loading with PLF. (b) Variation of SEC of mills with PLF.

average calorific value of coal from 3000 kcal/kg to near the design value of 4400 kcal/kg reduces the AP of mills by 0.07% of gross generation (12.8% of mill power).

3.1.6.1.2 Use of Lower Ash Content of Coal

The ash content of coal is increased over time due to opencast mining and depletion of good quality coal. The higher ash content of coal increases the AP and degrades the performance of the plant. The use of improved coal quality, that is, reduction in average ash content of coal from 50% to 20%, will reduce the AP of mills by 0.09% of gross generation (15.3% of mill power).

3.1.6.1.3 Use of Lower Moisture Content Coal

Generally in Indian coal, the moisture varies between 13% and 19%, which is higher compared with the design value of 10%. The reduction in AP of mills by operating the plant with reduced moisture content coal from the present value of 19% to the optimum value of 10% will be 0.03% of gross generation (5.1% of mill power).

3.1.6.1.4 Use of Beneficiated Coal

The beneficiating of coal is nothing but removing the impurities and ash from the coal by either wet or dry washing. This washing of coal reduces the raw coal size and the average ash content of coal from 52% to 31%, and increases the calorific value of coal from 3000 to 3550 kcal/kg (see Table 3.13). For the same power output, the coal flow is reduced by 8.3% and mill power input is reduced by 20.9%. The use of beneficiated coal has reduced the AP of mills by 0.14% of gross generation.

3.1.6.1.5 Use of Coal Blending

The use of imported blended coal in a typical power plant improved the net average calorific value from 2800 to 3940 kcal/kg, the coal flow is reduced by

10.3%, and the AP of mills is reduced by 21% (see Table 3.14). The use of coal blending reduced the AP of mills by 0.15% of gross generation.

3.1.6.1.6 Optimization of PC Fineness

The main aim of mills is to reduce the coal size to below 75 µm and 70% of the total PC must pass through a 200 size mesh (i.e., 75 µm size). The milling capacity and power consumption of mills depend on the PC fineness at the mill outlet, which is monitored by particle size analysis using sieves. The coal fineness is altered by adjusting the classifiers. In some power plants, dynamic classifiers are used to adjust coal fineness dynamically but these are not very popular. The higher fineness of coal increases the unburnt carbon in fly ash that escapes the firing zone faster, that is, the residence time for coal particles in the firing zone is much lower. On the other hand, if the coal fineness is lower, the mill capacity increases but the unburnt carbon increases in bottom ash in which the heavier particles fall below (as a bottom ash) before burning in the firing zone. Presently, there is no mechanism for maintaining equal coal flow in all the corners and consistent coal fineness. In one of the thermal power plants, online coal flow measuring (i.e., PC), balancing (control), and

TABLE 3.13

Performance Results of Mills with Beneficiated Coal for a Typical 210 MW Plant

S. No.	Particulars	Unit	Nonwashed Raw Coal	Washed Coal
01	Moisture	%	16.0	18.0
02	Ash content	%	52.0	31.0
03	Calorific value	kcal/kg	3000	3550
04	Total mill power	kW	1503.1	1189.5
05	Average mill loading	%	89.5	71.2
06	Mill differential pressure	kPa	3.44	2.20
07	Total coal flow	t/h	128.2	117.6
08	SEC of mills	kWh/t	11.72	10.11
09	Reduction in auxiliary power	kW (% of PL)	313.6 (0.14%)	

TABLE 3.14

Performance Results of Mills with Blended Coal at a Typical 210 MW Plant

S. No.	Particulars	Unit	Raw Coal	Blended Coal
01	Average calorific value	kcal/kg	3000	3940
02	Total coal flow	t/h	128.2	115.0
03	Total mill power	kW	1503.1	1187.1
04	SEC of mills	kWh/t	11.72	10.29
05	Reduction in total power	kW (% of PL)	316.0 (0.15 %)	

an online coal fineness measuring system for combustion optimization was introduced [40]. But at present, this system is costly and economically not cost-effective to implement in existing old power plants. A higher PC fineness at the mill outlet increases the AP as well as SEC of mills.

In many thermal power plants, the original equipment manufacturers (OEMs) have supplied Ni-hard steel bowls whose life is about 6000 h, but today most power plants are using hard chrome steel bowls and bull ring segments whose life has increased to about 7000–8000 h. It can be seen from Table 3.15 that the performance of mill E is poor because the particle size below 75 µm varies between 51.2% and 67.2%, which is far lower than the design value of 70% passing through 75 µm. Particles above 300 µm must be below 1.0% but in mill B, this is in the range of 2.5%–7.3%. The higher sized PC chokes up the coal pipes at bends and causes a situation of no coal flow in some corners, which shifts the fireball to a corner and disturbs the heat transfer in the boiler. The higher sized particles also increase the unburnt carbon in bottom ash (i.e., heat loss). Similarly, the performance of mill B and mill E is also poor. The performance of mill D is normal, where the particle size below 75 µm is around 70% and particle size above 300 µm is about 1.5%, but still this value should be brought down to below 1.0% for appropriate combustion. The performance of mill C is better, where the particle size below 75 µm is around 74.7%, but the SEC of the mill is marginally high and the particles above 300 µm is about 1.5%. This value can be brought down to below 1.0% for appropriate combustion.

3.1.6.1.7 Optimization of Raw Coal at Mill Inlet

The raw coal is received at the CHP and will be tippled at wagon tippler, and then the coal is fed to crushers to reduce the coal size to below 25 mm. In many power plants, the coal crushing will be done in two stages, that is, primary crushing where the coal size is reduced to 125 mm and then further

TABLE 3.15

Particle Size Analysis of Pulverized Coal Samples of Mills at Typical Power Plants

Mills	Mesh Size	Pulverized Coal Fineness (%)				
	Mesh/Corner	1	2	3	4	Average
Mill E	(+300 µm)	3.1	4.3	4.9	7.5	5.0
	(−75 µm)	51.2	67.2	55.6	57.3	57.8
Mill D	(+300 µm)	0.7	0.8	2.1	2.3	1.5
	(−75 µm)	71.8	80.9	53.6	72.2	69.6
Mill C	(+300 µm)	1.2	1.5	0.6	2.6	1.5
	(−75 µm)	77.1	74.2	76.1	71.3	74.7
Mill B	(+300 µm)	7.3	2.5	2.5	3.8	4.0
	(−75 µm)	55.5	61.1	61.2	73.2	62.8
Mill A	(+300 µm)	10.2	6.3	7.0	5.7	7.3
	(−75 µ)	52.3	75.1	68.7	66.1	65.6

TABLE 3.16

Performance Results of Crushers

S. No.	Particulars	Unit	Before Crusher	After Crusher
01	Coal weight below 25 mm size	kg (%)	62.5 (56%)	74.3 (80.8%)
02	Coal weight above 25 mm size	kg (%)	49.3 (44%)	17.7 (19.2%)
03	Total coal weight	kg (%)	111.8 (100%)	92.0 (100%)

crushing in a secondary crusher to reduce the coal size to below 25 mm. The crushers will generally be hammer mill type or beater type mills, whose SEC will be lower than that of coal mills. Larger raw coal at the mill inlet increases the mill power considerably. Larger raw coal at the mill inlet and foreign materials present in coal increases the mill outage and mill rejects in Raymond bowl mills. Generally for Raymond bowl mills, mill rejects of about 1.0% are allowed, but in many power stations, it is measured in the range of 1.0%–3.5%.

In a typical 210 MW power plant, the performance results of a crusher are measured by collecting raw coal samples at conveyor belts before the crusher and after the crusher simultaneously by stopping the crusher stream. Table 3.16 gives the performance results of the crusher. The total weight before the crusher was 111.8 kg, and the particle size below the 25 mm sieve size was 62.5 kg (56% of total coal) and above 25 mm was 49.3 kg (44%). After the crusher, the total weight was 92 kg, the raw coal size passing below 25 mm was 74.3 kg (80.8%), and the raw coal size above 25 mm was 19.2 kg (19.2%). The crusher increases the raw coal size below 25 mm from 56% to 80.8%. Reduction of raw coal size below 25 mm at mill inlet was from 19.2% to 10.0%, which reduces the AP of mills by 1.71% (0.01% of gross energy generation).

3.2 In-House LT AP

The in-house LT AP is the power used by in-house LT equipment like instrument air compressors, vacuum pumps, ESPs, control fluid pumps, stator water pumps, seal air fans, and so on. The in-house LT AP is fed from the boiler LT boards, turbine LT boards, and emergency LT boards. The LT power to these LT boards is fed from unit station transformers (USTs) with an LT power supply of 433/415 V.

The in-house LT AP is fed from auxiliary transformers in which the 6.6 kV voltage is stepped down to 0.433 kV. Generally, these auxiliary transformers are provided with off-load tap changers. The in-house LT AP (kW) is the difference between total in-house AP measured at UATs and the summation of AP measured at in-house HT auxiliary equipment, which is computed as

$$P_{\text{IAP-LT}} = \sum_{i=1}^{i=n} (P_{\text{UAT}})_i - P_{\text{IAP-HT}} \qquad (3.8)$$

where $P_{\text{IAP-HT}}$ is the in-house HT AP (kW), which is the summation of power used by HT auxiliary equipment and is computed as

$$P_{\text{IAP-HT}} = P_{\text{BFP}} + P_{\text{CEP}} + P_{\text{IDF}} + P_{\text{FDF}} + P_{\text{PAF}} + P_{\text{Mills}} \qquad (3.9)$$

where P_{BFP} is the average AP measured at BFPs (kW), P_{CEP} the average AP measured at CEPs (kW), P_{IDF} the average AP measured at ID fans (kW), P_{FDF} the average AP measured at FD fans (kW), P_{PAF} the average AP measured at PA fans (kW), and P_{Mills} the average AP measured at mills (kW).

The in-house LT AP varies between 1.52% of gross generation at MCR condition and 2.08% of gross generation at 70% PLF (see Figure 3.32). Operating the plant at 70% PLF increases the in-house LT AP by 0.56% of gross generation.

Table 3.17 gives the performance results of major in-house LT motors whose rating is above 25 kW in a typical 210 MW power plant. The observations and energy conservation measures for LT motors are discussed below:

1. The voltage at motor terminals plays a major role in the performance of motors. At reduced voltage, the motor losses will be more and torque delivered will be less, which will reduce the power output.

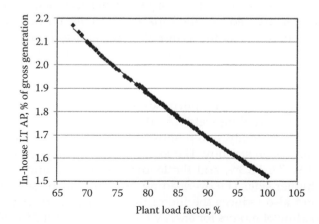

FIGURE 3.32
Variation of AP used for in-house LT motors with PLF.

TABLE 3.17

Performance Results of LT Motors at a Typical 210 MW Plant

S. No.	Particulars	Rating (kW)	Voltage (V)	Current (A)	Power Factor	Power (kW)	Load Factor (%)
01	Vacuum pump 2	72	425.4	95.0	0.804	56.3	70.4
02	ACW A	90	416.9	153.5	0.868	96.2	98.3
03	ACW C	90	426.6	141.6	0.895	93.7	95.7
04	ACW BP A	75	416.5	143.8	0.867	90.0	110.4
05	ACW BP B	75	415.5	140.2	0.812	82.0	100.5
06	ACW BP C	75	425.3	115.3	0.813	69.0	84.7
07	Seal air fan B	37	425.8	42.3	0.886	27.6	65.0
08	Seal air fan C	37	425.6	47.6	0.851	29.9	70.2
09	Stator water pump 2	55	417.2	85.2	0.889	54.7	91.6
10	HFO pump A	22	417.8	17.0	0.724	8.9	34.4

ACW, auxiliary cooling water; HFO, heavy furnace oil.

As per the manufacturer's recommendations, the motors will work with a voltage variation of ±10%. At many power plants, it was observed that the LT voltage at motor terminals is on the higher side, which strains the motor insulation and reduces the life of motors. The voltage at motor terminals can be maintained near the design value by altering the transformer taps of auxiliary transformers.

2. The voltage imbalance between the three phases at motor terminals creates a negative sequence current in the motor winding, causing heating of the motor winding that reduces the life of motor. The negative sequence current produces negative sequence torque in the motor that reduces the capacity of motor output. A voltage imbalance of 5% at motor terminals will reduce the motor capacity by 25% and increase the motor losses by 33%. But in thermal power plants, the voltage imbalance between the three phases will be much lower.

3. To meet auxiliary cooling purposes apart from condenser cooling, three auxiliary cooling water (ACW) pumps are installed in many power plants. Usually, two pumps are working continuously and a third pump is on standby. The load factor of these pumps is in the range of 95.7%–98.3%, and the loading of these pumps is slightly on the higher side. The overloading of motors increases the motor losses and also reduces the life of the motors. Therefore, the optimum loading of motors for energy efficiency is about 75%–85%.

4. In some power plants where the water source for auxiliary cooling is far away, ACW booster pumps are used to increase the pressure of

the ACW. In a typical power plant, these pumps were overloaded in the range of 100.5%–110.4%, which may be due to higher water flow or higher head or problems in pump internals. The overloading of motors strains the motor insulation and reduces the life of the motor. The overloading of LT motors may be avoided by overhauling and optimizing the water flow.

5. To provide the sealing air for mills to avoid the leakage of PC from the mills, sealing air with a higher pressure than the PA pressure is used inside the mills. Generally, three seal air fans are installed to cover the total seal air requirement for all mills, that is, two seal air fans are working continuously and third fan is on standby. The loading of these fans is in the range of 65.0%–70.2% and the motor loading is slightly on the lower side. The energy efficiency of motors working at a lower load factor is poor and causes higher energy loss. The power factor of lower loaded motors is also poor. From an energy efficiency point of view, the motors must be optimally loaded or better control techniques like use of VFD or intelligent motor controllers, and so on, can be used to enhance the energy efficiency of low loaded motors.

6. Two stator water pumps are installed to provide the stator water required for cooling of the generator stator, that is, one pump is working continuously and the other pump is on standby. The load factor of the stator water pump is 91.6%, which is slightly on the higher side.

7. To maintain the heavy furnace oil (HFO) pressure at oil guns, HFO pumps are operated to circulate the HFO from the tank to the burner and then back to the tank. One pump is working continuously and the other pump is on standby. The load factor of these pumps is on the lower side in the range of 30%–35%. It is suggested to install an intelligent motor controller for low loaded motors. This controller will sense the load and adjust the voltage and flux inside the motor to provide the required magnetizing current. This reduces magnetic losses in the motor and reduces the current at motor terminals. The power factor of the motor improves even at low load operation with this type of controller. This reduces the energy consumption for the above low loaded motors by about 10.5 MWh/month. The anticipated investment for this type of controller is $0.2 million and the simple payback period is 35 months.

8. The motor winding failure rate is higher in many power plants because of the use of same motor after several rewindings. The motors are rewound frequently. The rewinding of motors reduces the insulation property of the motors. It is suggested to measure the no-load current of all the LT motors and maintain the records. If the no-load current of motor increases (about 25%–30%), it is better to replace the entire motor rather than rewinding. The rewinding

of motors may be allowed only two to three times. If the motor is rewound three times, it is better to replace this motor with an energy-efficient motor.

3.3 Conclusions

The total in-house AP varies between 6.5% and 8.3% of gross power generation including excitation power and losses in GT. The total in-house AP forms about 72.5%–81.3% of the total AP of the plant. The major in-house HT equipment are powered with 6.6 kV for a 210 MW plant or 11 kV for a 500 MW plant. The average in-house HT AP forms about 5.6% of gross generation for a 210 MW plant. The loading of these HT motors directly depends on the plant load on the individual units. The average AP used by various components at MCR condition is as follows: BFPs, 2.44%; CEPs, 0.22%; ID fans, 1.12%; FD fans, 0.215%, PA fans, 0.93%; mills, 0.66%; and in-house LT equipment, 1.52%.

Some of the energy conservation measures are the following:

1. Reducing the pressure drop of the FW across the FRS from an average value of 0.35–0.10 MPa by adopting a three-element control technique will reduce the AP of the BFP by 1.7%; replacement of the BFP impeller (cartridge) will reduce the AP of a BFP by 13.1%; and replacement of the valve seat of the RC valve for a BFP will reduce the AP of a BFP by 8.1%. The energy conservation potential for a BFP is 0.56% of gross generation.

2. Reducing the SSC from 3.3 to 3.0 t/MW by improving the turbine heat rate of a turbine will reduce the AP of the CEP by 12.5%, and reducing the DM water makeup from 3.5% to 1.0% will reduce the AP of the CEP by 1.8%. The energy conservation potential for the CEP is 0.03% of gross generation.

3. Using beneficiated coal will reduce the AP of ID fans by 8.8%, FD fans by 3.9%, PA fans by 19.5%, and mills by 20.9%. The average reduction in AP by using beneficiated coal is 0.43% of gross generation.

4. Blending of Indian coal with imported coal (proportion 80:20) and adopting a tier blending technique reduces the AP of mills by 21.0%, PA fans by 19.3%, FD fans by 2.0%, and ID fans by 7.5%. The average reduction in AP by blending of Indian coal with imported coal is 0.41% of gross generation.

5. Control of excess air by reducing the oxygen content at the APH inlet from an average value of 4.3% to 3.5% will reduce the AP of ID fans by 2%, FD fans by 5.1%, and PA fans by 1.9%. The average reduction in AP by control of excess air is 0.05% of gross generation.

6. Reducing the air leakage through the APH by controlling the rise in oxygen content to about 1% will reduce the AP of ID fans by 5.4%, FD fans by 10.4%, and PA fans by 2.7%. The average reduction in AP by control of air leakage through the APH is 0.11% of gross generation.

7. Reducing the air leakage through the flue-gas duct by monitoring and controlling the rise in oxygen content to about 1% between the APH outlet and ID fan inlet will reduce the AP of ID fans by 4.2% (0.04% of gross generation).

8. Reduction of the flue-gas pressure drop across the APH from 1.25 to 1.02 kPa reduces the AP of ID fans by 5.2% (0.05% of gross generation) and reducing the flue-gas pressure drop across the ESP from 0.90 kPa to 0.44 kPa reduces the AP of ID fans by 8% (0.09% of gross generation).

9. Reduction of the SA pressure drop across the APH from 0.88 to 0.86 kPa reduces the AP of FD fans by 1.7% (0.004% of gross generation).

10. Reduction of the PA pressure drop across the APH from 2.33 to 0.42 kPa reduces the AP of PA fans by 9.7% (0.09% of gross generation).

11. Optimizing the PA fan discharge pressure from 9.4 to 8.0 kPa will reduce the AP of PA fans by 13.7% (0.13% of gross generation).

The implementation of energy conservation measures reduces the average AP of in-house AP by 1.6% of gross generation.

4

Energy Conservation of Common Auxiliary Power Equipment in Power Plant Processes

Rajashekar P. Mandi

School of Electrical and Electronics Engineering, REVA University, Bangalore, India

Udaykumar R. Yaragatti

Department of Electrical and Electronics Engineering, National Institute of Technology Karnataka (NITK), Surathkal, India

4.1 Introduction

Common auxiliary power (AP), which is also known as station AP, is the power used to drive the auxiliary equipment that is utilized by more than one unit in the entire power station. The running of common auxiliary equipment does not depend on the operation of a particular unit. So, the change in plant load on a particular unit will not directly vary the power used by common auxiliary equipment. The power to this equipment is fed from station transformers (STs). Common AP is subclassified based on voltage level into high tension (HT) AP and low tension (LT) AP; it is also subclassified based on utilities such as the coal-handling plant (CHP), ash-handling plant (AHP), air compressor system, and water treatment plant (WTP). The average total common AP varies between 1.1% and 2.7% of gross generation. The average common AP of a typical 210 MW power plant is about 1.78% of gross power generation (17.4% of total AP). The common (outlying) AP P_{CAP} (MW) is computed as follows:

$$P_{CAP} = \frac{\sum_{j=1}^{j=p} (P_{ST})_j}{PL \times n} \times 100 \tag{4.1}$$

where P_{ST} is the measured power (MW) of ST, PL is the plant load (MW), n is the number of units, and p is the number of STs.

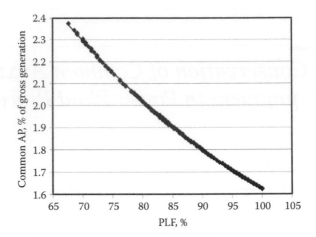

FIGURE 4.1
Variation of common auxiliary power (AP) with plant load factor (PLF).

The common AP includes power used by both common (outlying) HT and LT auxiliary equipment.

Figure 4.1 shows the variation of common AP measured during the performance test at power units. It can be seen in the figure that as the plant load factor (PLF) increases, the common AP increases in magnitude and the common AP as a percentage of gross generation decreases. The common AP varies between 1.62% of gross generation at maximum continuous rating (MCR) condition and 2.29% at 70% PLF.

4.2 Coal-Handling Plant

The AP of the CHP is in the range of 0.12%–0.22% of gross generation and varies widely depending on the topography of the area for the CHP. With an increase in monthly average PLF, the AP (as a percentage of gross generation) of the CHP decreases (see Figure 4.2). As the average monthly coal flow increases, the AP (as a percentage of gross generation) decreases (see Figure 4.3). The mode of sourcing of coal from collieries to plants depends on the location of the plant with respect to the coal mines. If the plants are away from the mines, the coal is transported by railway wagons. The coal is received in railway wagons and is tippled by either automatic or manual wagon tipplers. In the case of pithead power plants, the coal is supplied by a merry-go-round (MGR) system where the coal is unloaded by bottom discharge wagons onto the track hopper [41]. In some pithead power plants, the coal is supplied through a ropeway system. For coastal power plants, the coal is supplied through ships to ports and from ports coal is transported through belt or pipe to the plant. At major power plants, the coal is supplied by railway wagons. Figure 4.4 shows the

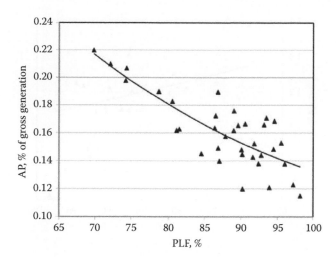

FIGURE 4.2
Variation of AP with PLF of coal-handling plant (CHP).

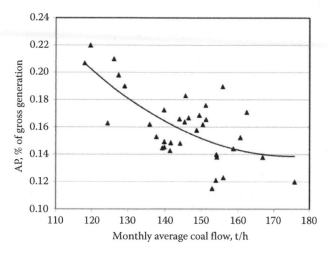

FIGURE 4.3
Variation of AP with coal flow.

variation of specific energy consumption (SEC) with PLF of a typical power plant. The average monthly SEC varies between 0.91 and 1.39 kWh/t of coal handled in the CHP. Figure 4.5 shows the variation of SEC with monthly average coal flow. As the coal flow increases, the SEC decreases.

The design coal flow capacity of the CHP is 750–1800 t/h, depending on the number of units and unit sizes, but the utilization factor is in the range of 50%–70% depending on the condition of the coal (wet or dry). The conveyor belts are designed for 1000–1500 t/h capacity, but their utilization factor is about 40%–60%.

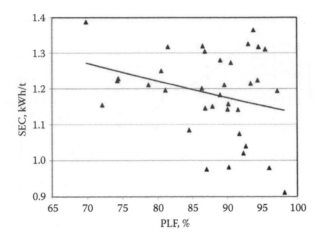

FIGURE 4.4
Variation of specific energy consumption (SEC) with PLF.

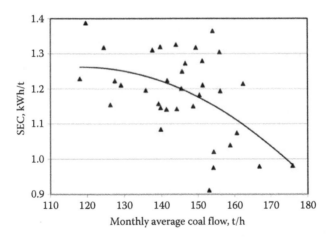

FIGURE 4.5
Variation of SEC with coal flow.

Figure 4.6 shows the schematic diagram of the CHP of a typical power plant. The design capacity of a wagon tippler is generally of about 90 t. The coal is received by a railway in coal rakes consisting of 58–60 BOXN type wagons where each wagon consists of a payload of about 55–60 t. The coal wagons are tippled by rotary type wagon tipplers. The tipplers are positioned either manually or by beetle chargers or side arm chargers. Many power plants practice wagon positioning manually, which takes about 2–4 min, but regular locomotive engines must be used whose energy consumption is higher. In beetle charger wagon positioning, about 10–15 wagons are grouped together and are moved by an electrically-operated system. Similarly, in a side arm charger, about 18–20 wagons are grouped, and the side arm is electrically

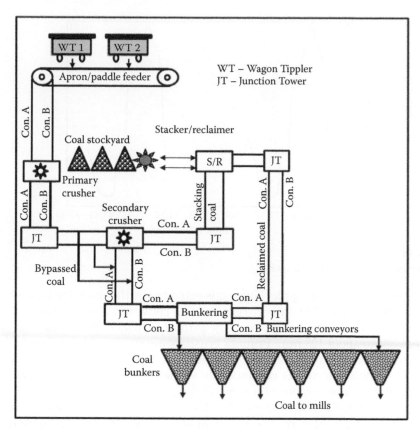

FIGURE 4.6
Schematic of CHP.

operated to move the wagons one by one. The time taken by a beetle charger, as well as a side arm charger, is comparatively less (about 30–80 s), but the maintenance cost is slightly higher and the net energy consumption for wagon positioning is less compared with manual positioning.

The average tippling time taken by wagon tipplers varies between 60 and 80 s for the onward direction and the same time for return of the wagon. The total time for clearing one wagon depends on the positioning of the wagon, coal clearing in the chute, and loading of belts (i.e., paddle feeders, apron feeders). The increased time for wagon clearing increases the demurrage charges for wagons to the railway department. Generally, the total weight of coal in each wagon is in the range of 55–70 t. Figure 4.7 shows the variation of power during tippling of a wagon in a typical power plant. The peak power during tippling for a wagon varies in the range of 90–110 kW. The energy used for tippling in a forward motion is 1.2–1.8 kWh and the returning energy in the reverse direction is 0.25–0.42 kWh. The average total energy consumption for tippling one wagon is 1.5–2.2 kWh, and the average SEC varies from 25×10^{-3} to 55×10^{-3} kWh/t of coal.

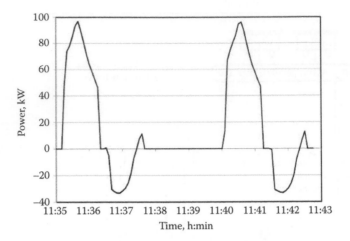

FIGURE 4.7
Variation of power of wagon tippler.

The coal from wagon tippler hoppers is extracted by the apron feeders, which then transfer the coal to an underground belt conveyor. The coal then moves on to various conveyor belts toward the primary crushers where the coal size is reduced to about 125 mm. Primary crushers are run by either LT motors or HT motors of rating 200–300 kW. Generally, two primary crushers are provided, but only one works continuously while the other is kept on standby (100%). But in an emergency (in some of the power plants) both streams can be operated simultaneously in parallel, that is, one stream for stacking and the other stream for bunkering. For each primary crusher, there is a provision for bypassing the coal. Generally, primary crushers are either Bradford breaker or jaw-type roller crushers where the output coal size is 100–150 mm (see Table 4.1). But in some power plants where the coal input size is very big, two stages of primary crushing are used. In this type of plant, double-roller crushers are used to reduce the coal size to about 300 mm and then, using other primary crushers, the coal size is reduced to 100–150 mm. The majority of the power plants use a single primary crushing mechanism. The average SEC for primary crushing is 0.07–0.17 kWh/t of coal. In the majority of the power plants, the loading of primary crushers is 20%–30%, which causes higher SEC.

The coal from the primary crusher goes to a number of conveyors and then to secondary crushers where the coal size is reduced to 25 mm. Secondary crushers are run by HT motors of rating 500–1200 kW depending on the size of the plant. The secondary crushers are either impact beater type or hammer mills or ring granulator type. For secondary crushers, hammer mills and ring granulators are commonly used. The SEC for a secondary crusher is 0.35–0.78 kWh/t of coal and is higher than the primary crusher because the input and output coal size is smaller for a secondary crusher (i.e., more grinding is required for a secondary crusher as compared to a primary crusher). Generally, for each stage, two secondary crushers are provided, with only

TABLE 4.1

Performance Results of Crushers

		Output Coal	SEC (kWh/t)	
S. No.	Particulars	Size (mm)	Design	Operating
01	Ring granulator	25	0.61	0.72–0.78
02	Hammer mill	25	0.50	0.36–0.61
03	Impact beater	25	0.70	0.65–0.70
04	Jaw-type roller crusher	100	0.10	0.07–0.09
05	Bradford breaker	150	0.25	0.14–0.17
06	Double-roller-type crusher	300	0.23	0.13–0.24

Note: SEC, specific energy consumption.

one crusher working while the other crusher kept on standby (100%). But in some power plants, both streams can be operated simultaneously in parallel, that is, one stream for stacking and the other for bunkering. In some power plants, only a single crushing is used, that is, only a secondary crusher is used for crushing without a primary crusher in series where the SEC for the secondary crusher increases but the net SEC for crushing coal decreases because of the absence of a primary crusher. For each secondary crusher, either a vibratory screen or roller screen is installed to bypass the coal below 25 mm. At the outlet of the secondary crushers, a 25-mm screen is provided to allow only the coal below 25 mm to exit from the crusher to the mills. The loading of these secondary crushers varies widely from 20% to 60%. Generally, in many power plants, the coal size at the crusher inlet below 25 mm varies between 30% and 50%, and only about 50%–80% of the coal is being crushed in secondary crushers where bypass screens are provided. In some power plants, the secondary crusher output screens are removed to avoid choking of coal that allows coal larger than 25 mm. If coal size above 25 mm is more prevalent at the crusher output, the SEC of mills increases considerably because the grinding principle in mills is different from crushers. A coal particle size analysis is carried out for a typical power plant to evaluate the performance of crushers. In this, the raw coal is collected from the conveyor belt before and after the crusher, simultaneously. The analysis results are given in Table 4.2. It can be seen from the table that coal smaller than 25 mm at the crusher inlet is 57.0% and this is increased to 66.2% after the crusher, which indicates that the secondary crusher had crushed only 9.2% of the coal. Coal larger than 25 mm before the crusher is about 43.0% and is reduced to about 33.8%. This higher percentage of raw coal size above 25 mm increases the AP of mills.

The coal from the secondary crusher moves on to either bunkering (coal fed to unit bunkers) or stacking, that is, storing the crushed coal for future use. The coal moves through a number of belt conveyors for direct bunkering. The specific energy consumption for bunkering varies widely from 0.6 to 1.6 kWh/t

TABLE 4.2

Performance Results of Crushers

S. No.	Particulars	Before Crusher [kg (%)]	After Crusher [kg (%)]
01	Coal weight below 25 mm size	62.9 (57.0)	32.7 (66.2)
02	Coal weight above 25 mm size	47.5 (43.0)	16.7 (33.8)
03	Total coal weight	110.4 (100)	49.4 (100)

of coal handled because of the different topography in different plants. The average AP for bunkering is 0.05%–0.09% of gross generation, and the loading of conveyor belts is 20%–60%. The SEC for stacking the coal varies widely from 0.5 to 1.2 kWh/t of coal handled, and the average AP is 0.02%–0.07% of gross generation. The stacked coal is used for bunkering through the reclaimer whenever coal wagons are not available. The SEC for reclaiming of coal varies widely from 0.2 to 0.6 kWh/t of coal handled, and the average AP used for reclaiming is 0.01%–0.03% of gross generation. Table 4.3 shows the performance results of crushers and conveyor belts in a typical 210 MW power plant.

4.2.1 Energy Conservation Measures

1. Wagon positioning can be done by using a beetle charger, which reduces the time for positioning of the wagon as well as the oil consumption of the loco engine.
2. Use of a vibratory bypass screen for a secondary crusher saves the energy consumption of secondary crushers by about 20%–30%.
3. Use of a secondary crusher output screen to allow only coal below 25 mm to mills reduces the SEC of mills considerably.
4. In many power plants, the crusher motor load factor is 5%–45%, which is very poor, and the load factor of conveyor motors is 5%–50%, which is also on the lower side. Operating the motors at a lower load factor reduces the power factor as well as motor efficiency, which increases the losses. Therefore improving the loading of crushers and conveyor belts reduces the energy consumption.
5. The SEC of crushers is 0.02–0.49 kWh/t of coal handled and depends on the loading of crushers, which again directly depends on the performance of the conveyor system, wagon tippler, availability of coal, bunkering of coal, and so forth. Generally, a crusher's coal loading is very low, 10%–40%, which creates a higher SEC. To reduce the AP of crushers, the crushers must be loaded above 70%.
6. The SEC of conveyor belts is 0.02–0.20 kWh/t of coal handled and directly depends on the coal handled. The SEC of conveyors is high due to underloading of conveyor belts, that is, the conveyor systems

TABLE 4.3

Performance Results of Crushers and Conveyor Belts in a Coal-Handling Plant (CHP)

S. No.	Particulars	Rating (kW)	Voltage (V)	Current (A)	Power Factor	Power (kW)	Load Factor (%)	SEC (kWh/t)
				HT				
1	Secondary crusher A	737	6703	45.5	0.690	364.1	43.5	0.485
2	Secondary crusher B	737	6594	46.7	0.667	355.9	42.5	0.475
3	Secondary crusher D	750	6754	41.4	0.156	75.7	8.9	0.101
4	Primary crusher C	225	6770	11.6	0.414	56.5	22.1	0.075
5	Primary crusher D	225	6773	12.8	0.470	70.4	27.6	0.094
				LT				
6	Primary crusher A1	180	442.2	67.9	0.308	16.0	7.8	0.021
7	Primary crusher B1	180	424.6	84.3	0.692	42.9	21.0	0.057
				HT				
8	Conveyor 101A	213	6673	13.6	0.712	112.0	46.3	0.149
9	Conveyor 102A	213	6693	7.9	0.452	41.3	17.1	0.055
10	Conveyor 103A	315	6588	16.6	0.765	145.0	40.5	0.193
11	Conveyor 103B	315	6690	15.9	0.743	136.7	38.2	0.182
12	Conveyor 104A	269	6642	11.2	0.634	81.4	26.6	0.109
13	Conveyor 104B	373	6792	16.4	0.668	128.8	30.4	0.172
14	Conveyor 106A	250	6644	11.1	0.635	80.9	28.5	0.108
15	Conveyor 106B	250	6650	13.8	0.764	121.1	42.6	0.162

(Continued)

TABLE 4.3 (*Continued*)

Performance Results of Crushers and Conveyor Belts in a Coal-Handling Plant (CHP)

S. No.	Particulars	Rating (kW)	Voltage (V)	Current (A)	Power Factor	Power (kW)	Load Factor (%)	SEC (kWh/t)
16	Conveyor 107A	280	6644	10.5	0.444	53.5	16.8	0.071
17	Conveyor 107B	280	6644	15.0	0.744	128.7	40.4	0.172
18	Conveyor 112B	298	6762	16.1	0.720	135.4	40.0	0.181
19	Conveyor 113A	220	6776	14.8	0.695	120.7	48.3	0.161
20	Conveyor 115B	285	6798	17.4	0.709	145.2	44.8	0.194
				LT				
21	Conveyor 101B	110	441.9	84.7	0.811	52.6	42.0	0.070
22	Conveyor 102B	110	425.8	92.8	0.836	57.2	45.8	0.076
23	Conveyor 105	37	436.3	38.0	0.549	15.8	37.5	0.021
24	Conveyor 107	110	451.5	100.0	0.348	27.2	21.8	0.036
25	Conveyor 114	45	426.2	38.3	0.745	21.1	41.2	0.028
26	Conveyor 116	55	421.4	31.2	0.722	16.4	26.3	0.022
27	Conveyor 108A	120	422.1	82.5	0.748	45.1	33.1	0.060
28	Conveyor 108B	120	434.8	96.1	0.773	56.0	41.1	0.075
29	Conveyor 121	80	448.1	53.5	0.567	23.5	25.9	0.031
30	Conveyor 125	55	436.6	44.8	0.661	22.4	35.8	0.030
31	Conveyor 130B	100	434.7	64.5	0.645	31.3	27.6	0.042
32	Conveyor 131B	110	439.8	74.0	0.600	33.8	27.1	0.045

Note: HT, high tension; LT, low tension.

in many power plants are designed with a capacity of 1000–1500 t/h coal flow but are loaded very low, 300–700 t/h, because of fear of damage to the belts, spilling of coal from belts, and so forth.

7. Conveyor belts are placed on top of rollers to reduce the frictional losses. But in many power plants, it was observed that rollers are not free running and stuck, causing higher frictional loss. These rollers should be free running to reduce the frictional losses for conveyor belts, which in turn will reduce the energy consumption of belts.

8. In many power plants, the coal belts are tilted to one side, which spills the coal. The spilled coal accumulates on either side of the conveyor belts, which causes obstruction in belt movement. The periodic inspection of conveyor belts, adjusting the belts, and clearing the coal on either side of conveyor belts help in reduction of energy consumption.

4.3 Ash-Handling Plant

The energy used by the AHP is 0.10%–0.25% of gross generation. The energy consumption varies for different plants depending on the ash dyke. The ash-handling system consists of mainly ash slurry pumps, high-pressure water pumps (HPWPs) and low-pressure water pumps (LPWPs), and so forth.

The coal-fired thermal power plant generates a large amount of ash, which is of great concern, and disposal of this ash is a major problem for thermal power generation. The average specific ash generation is 0.2–0.3 t/MW but depends on the ash content and calorific value of the coal used. The evacuation, collection, and disposal of ash in an environmentally friendly manner is a substantial task. The ash produced in coal-fired power plants is in two forms, that is, bottom ash (coarse ash), which is collected at the furnace bottom in wet form, and fly ash (fine ash), collected at different hoppers such as air preheaters (APHs), the economizer, and flue-gas ducts and finally at the electrostatic precipitator (ESP). The majority of fine fly ash that is entrained in the flue gas is collected in the ESP. The composition of bottom ash and fly ash depends on the size of the pulverized coal burnt in the furnace. If the pulverized coal fineness is higher, the ash escapes along with flue gas, which increases the fly ash component in the ESP. If the pulverized coal fineness is less, that is, there are more coarse particles, the majority of ash falls down as bottom ash in the furnace bottom. Generally, the ratio of bottom to fly ash is 20:80.

The water required for the ash-handling system is met by the blowdown of the cooling tower (CT), and the raw water is also used for mixing with ash slurry. The ash water system consists of HPWPs, LPWPs, seal water pumps, and economizer ash water pumps.

4.3.1 Bottom Ash

Bottom ash is collected at the bottom of the furnace where water is filled to prevent air ingress into the furnace. Sometimes the bottom ash is in the form of clinker and is grinded by the clinker grinder. The ash at the exit of the clinker grinder is mixed with the water to form slurry, and this slurry is pumped by jet pump to a common ash pond. At the slurry pond, the bottom ash slurry is mixed with fly ash slurry.

Generally, in a typical power plant, two clinker grinders are installed, with only one grinder working and the other grinder kept on standby. The clinker grinder is a single/double roll type, and its speed is about 35–40 rpm. It is provided with a reversing mechanism to reverse the direction of the grinder rolls. It crushes large clinkers to suitable size, that is, below 25 mm for transportation through the pipeline. The coarse ash is collected at the economizer hopper bottom, and is transferred to the bottom ash hopper top by means of an adequately sized sloping pipe that uses gravity to transfer bottom ash. If the bottom ash hopper capacity is not sufficient, then economizer hopper ash is transported separately to the main slurry tank.

Bottom ash high pressure (BAHP) water pumps are used to extract bottom ash intermittently and sequentially. In the case of a jet pump system, BAHP pumps supply water for jet pumps, bottom ash hopper flushing, seal trough, gate housing flushing, and so forth. Bottom ash low pressure (BALP) pumps supply water for refractory cooling, bottom ash hopper cooling water to maintain the hopper water, bottom ash hopper fill and makeup, seal trough makeup/fill, slurry sump hopper makeup water, and so forth. In a typical power plant, three bottom ash LP water pumps of 55 kW are installed, of which two pumps work continuously for two units of 210 MW (see Table 4.4). The load factor of these water pumps is 53.3%–59.6%. The loading of the pump motor on the lower side may be due to less water flow. The SEC for bottom ash water pumps is 2.1–3.4 kWh/t of bottom ash handled (0.14–0.21 kW/MW of plant load).

TABLE 4.4

Performance Results of Pump Motors at the Bottom Ash Pump House

S. No.	Particulars	Rating (kW)	Voltage (V)	Current (A)	Power Factor	Power (kW)	Load Factor (%)
01	Bottom ash water pump 1	55	414.6	58.7	0.88	37.2	59.6
02	Bottom ash water pump 2	55	416.0	58.3	0.86	36.3	58.1
03	Bottom ash water pump 3	55	418.5	53.4	0.86	33.3	53.3
04	Clinker grinder	7.5	417.7	6.7	0.29	1.4	16.3

4.3.2 Fly Ash

Fly ash is evacuated and transported in two ways, that is, using dry and wet systems. In a dry system, evacuation is implemented in two stages, that is, from ESP collection hoppers (where a large volume of fly ash is collected) to the intermediate surge hoppers by vacuum pumps and then to storage silos by transport air compressors (pneumatic pressure conveying). At the APH and duct hoppers, ash is conveyed pneumatically to an intermediate surge hopper and then to a silo. The dry ash from the silo is collected in tankers and transported to nearby industries for utilization.

In some power plants, the fly ash is disposed by a wet system where the fly ash is mixed with water at the ESP hopper bottom and then transported to the main slurry tank. Fly ash HPWPs supply water to wetting heads, air washers, fly ash slurry, trench jetting, combined ash slurry sump makeup, combined ash slurry sump agitation, and so forth. Seal water pumps are provided for gland sealing of slurry pumps, vacuum pumps, and cooling compressors and the sealing water requirements of clinker grinders.

In a typical power plant, three fly ash LPWPs of 22 kW are installed, of which two pumps work continuously for two units of 210 MW (see Table 4.5). The load factor of these water pumps is 81.8%–107.8%. FALPWP 2 and 3 both are possibly overloaded due to higher water flow through the pumps. The overloading of pump motors strains the motor insulation and reduces the life of the motor considerably. Four fly ash HPWPs of 315 kW are installed, of which two pumps work continuously and the other two are kept on standby. The

TABLE 4.5

Performance Results of Pump Motors at a Fly Ash Pump House

S. No.	Particulars	Rating (kW)	Voltage (V)	Current (A)	Power Factor	Power (kW)	Load Factor (%)
01	Fly ash LPWP 1	22	418.5	33.9	0.83	20.5	81.8
02	Fly ash LPWP 2	22	413.2	41.3	0.86	25.3	101.4
03	Fly ash LPWP 3	22	413.9	41.8	0.90	26.8	107.4
04	Fly ash HPWP 1	315	6520.0	27.6	0.86	269.5	75.3
05	Fly ash HPWP 2	315	6533.3	26.6	0.87	263.4	73.6
06	Fly ash HPWP 3	315	6556.7	31.4	0.84	298.6	83.4
07	Fly ash HPWP 4	315	6556.7	27.0	0.88	270.8	75.6

Note: HPWP, high-pressure water pump; LPWP, low-pressure water pump.

load factor of these fly ash HPWP pump motors is 73.6%–83.4%, and the loading of these motors is normal. The SEC for fly ash water pumps varies widely from 4.5 to 5.7 kWh/t of fly ash handled (0.4–0.8 kW/MW of plant load).

To conserve water used in wet ash disposal, an ash water recovery system is provided to recirculate the decanted water from the ash pond and reuse it for ash-handling purposes.

4.3.3 Slurry Pumps

The total ash slurry is disposed to an ash dyke (which is located away from the power plant) through ash slurry pumps. A number of ash slurry pumps are connected in series to overcome the frictional loss in the piping system depending on the distance of the ash dyke from the plant. As the distance increases, the number of pumps in series increases to provide the sufficient head for disposal of ash slurry into the ash dyke. At a typical 210 MW plant, three pumps (180 kW) are connected in series to dispose the slurry into an ash dyke 7 km away from the plant (see Table 4.6). There are three groups of slurry pumps, with one group in service continuously and the other two groups kept on standby. The SEC for slurry pumps is 0.11–0.13 kWh/t of ash slurry per kilometer.

4.3.4 Energy Conservation

Generally, in power plants, the water to ash ratio is not monitored regularly, and is 5:1 to 15:1 in many plants. The increased water to ash ratio increases the pumping power, and the increase in the ratio to 15:1 increases the power by 75% compared with a 5:1 ratio. A too low water to ash ratio reduces the viscosity of the slurry, which leads to increase in frictional loss in the piping

TABLE 4.6

Performance Results of Ash Slurry Pump Motors at a Fly Ash Pump House

S. No.	Particulars	Rating (kW)	Voltage (V)	Current (A)	Power Factor	Power (kW)	Load Factor (%)
01	ASP A1	220	6496.7	23.9	0.78	209.1	83.6
02	ASP A2	220	6496.7	21.3	0.81	193.3	77.3
03	ASP A3	220	6506.7	15.3	0.70	120.2	48.1
04	ASP B1	220	6490.0	19.1	0.83	177.5	71.0
05	ASP B2	220	6506.7	19.5	0.84	183.9	73.6
06	ASP B3	220	6507.8	19.4	0.85	185.9	74.3
07	ASP C1	220	6536.7	22.0	0.78	193.9	77.5
08	ASP C2	220	6540.0	15.1	0.80	136.4	54.6
09	ASP C3	220	6540.0	21.7	0.81	198.9	79.6

Note: ASP, ash slurry pump.

system and causes erosion of the pump impeller. Therefore, the water to ash ratio must be optimized.

4.4 Circulating Water Plant

In a coal-fired thermal power plant, the steam exhaust from the LP turbine is converted from steam vapor to condensate in the condenser with the help of circulating water. The circulating water is also used for secondary cooling of the boiler and turbine auxiliaries.

Generally, in a thermal power plant, there are two types of circulating water cooling systems, that is, one is an open cycle cooling water system (once through the system) where the cooling water is passed once through the condenser, and this system is implemented where a large source of water is available, such as sea water for coastal power plants or where abundant river water is available [42]. The other is a closed cycle system (recirculated system) where the circulating water is recirculated after cooling the hot water in the CT and adding less makeup water than the first method stated above [43]. Initially, the once-through cooling system was the favored approach. The large size of the source water guarantees cold cooling water temperatures with very modest seasonal variation. With lower makeup (intake) flows and lower discharge (CT blowdown) flows, recirculated cooling systems were increasingly adopted in India.

Generally, two types of CTs are adopted in a closed cycle system, that is, an induced draft type or natural draft type depending on the techno-economics involving capital cost, operating expenses, and consideration of site-specific issues. Induced draft cooling towers (IDCTs) are preferred for power plants located near pithead power plants where operating expenses are lower (the overall power generation cost is less). On the other hand, natural draft cooling towers (NDCTs) are preferred for power plants located at load centers (far off from the pithead where fuel transportation cost is high), as these do not involve any rotating equipment, thus saving AP for CT fans. Air flow rate through the NDCT depends on the density difference between ambient air and the relative hot and humid light air inside the tower. For sites with a considerable duration of high summer ambient temperatures coupled with low relative humidity values, an adequate density difference would not be available for proper design and operational performance of an NDCT. For such sites, an IDCT should be preferred over an NDCT.

In the case of a once-through system, a desilting arrangement and traveling water screens of appropriate mesh size should be provided at the water intake to prevent debris and biological species in the incoming water from entering and mixing with cooling water. In case of a sea water–based cooling water system, debris filters of appropriate mesh size are provided at the upstream section of the condenser for further removing debris from the cooling water and thus reducing fouling of the condenser tubes.

4.4.1 Circulating Water Pumps

In a typical 2 × 210 MW power plant, there are six circulating water (CW) pumps of 1650 kW for both units (three pumps for each unit), and the CTs are of the natural draft type. Two pumps for each unit are in service continuously and the third pump is kept on standby. For auxiliary cooling purposes, there are four auxiliary cooling water (ACW) pumps of 630 kW (two pumps for each unit). One pump for each unit is in service continuously and the second pump is kept on standby. The auxiliary power used by a CW pumps is 1.2%–1.3% of gross energy generation and the AP of an ACW pump is 0.19%–0.21% of gross energy generation. The performance results of CW and ACW pumps are presented in Tables 4.7 and 4.8, respectively, for a typical 210 MW power plant.

- The CW flow at pumps varies between 5.26 and 5.42 m³/s, which is higher (by 16%) than the design value of 4.67 m³/s, possibly due to lower discharge pressure (158–162 kPa compared with a design value of 252 kPa net head).

TABLE 4.7

Performance Results of CW Pumps at a Typical Power Plant

S. No.	Particulars	Unit	Design	CWP A	CWP B	CWP C
01	CW pump suction pressure	kPa	–	−39.7	−39.7	−39.7
02	CWP discharge pressure	kPa	247.2[a]	161.9	157.9	157.9
03	CW flow	m³/s	4.67	5.42	5.38	5.26
04	Electrical power input	kW	1445	1629.9	1584.3	1641.8
05	Load factor of motor	%	82.3	89.0	86.3	93.6
06	Mechanical power output	kW	1154.5	1091.6	1062.7	1040.7
07	Combined efficiency	%	80.0	67.0	67.1	63.4
08	SEC	kWh/m³ of CW flow	0.09	0.08	0.08	0.09

[a] *Total net head.*

TABLE 4.8

Performance Results of Auxiliary Cooling Water (ACW) Pumps at a Typical Power Plant

S. No.	Particulars	Unit	Design	ACWP A	ACWP B
01	ACW pump suction pressure	kPa	–	−35.6	−43.5
02	Pump discharge Pressure	kPa	688.7[a]	545.4	564.1
03	ACW flow	m³/s	0.56	0.59	0.61
04	Electrical power input	kW	480.0	489.3	528.4
05	Load factor of motor	%	84.9	71.8	77.8
06	Mechanical power	kW	382.6	323.1	342.6
07	Combined efficiency	%	80.0	66.0	64.8
08	SEC	kWh/m³ of ACW flow	0.24	0.23	0.24

Note: ACWP = auxiliary cooling water pump.
[a] *Total net head.*

- CW pump combined efficiency is on the lower side at 63.4%–67.0% compared with a design value of 80%. The energy loss due to deviation in combined efficiency of CW pumps is 0.3% of gross generation.
- The SEC of CWP pumps is computed in the range of 0.08–0.09 kWh/m³ of CW flow and is on par with the design value.

4.4.1.1 Energy Conservation Measures

- Pumps can be coated with surface coating, which reduces pump internal losses, improves pump efficiency, and reduces the SEC of pumps. The anticipated energy savings is 111 MWh/month for a 2 × 210 MW plant, and the approximate investment for three pumps is US$20,000. The simple payback period is 4 months.
- The pressure drop across a condenser tube on the CW side was 32–40 kPa, which is lower than the optimal value of 40 kPa. The increased pressure drop across condenser tubes (on the CW side) increases the AP of CW pumps. The remedial measure is to reduce the pressure drop in the condenser tubes by cleaning (descaling) the condenser tubes using a high-pressure water jet periodically or acid cleaning.
- The water flow at the ACW pumps varies between 0.59 and 0.61 m³/s and is high compared with the design value of 0.56 m³/h, possibly due to a lower pump discharge pressure (545–564 kPa compared with a design value of 689 kPa net head).

- The ACW pump combined efficiency is 64.8%–66.0% and is on the lower side compared with the design value of 80%. The energy loss due to deviation in the combined efficiency of ACW pumps is 0.05% of gross generation.
- The SEC of ACW pumps is 0.23–0.24 kWh/m³ of ACW flow and is on par with the design value.

In another typical 210 MW power plant where IDCT fans were installed, two CW pumps of 1400 kW were working continuously. It was provided with one IDCT with 10 CT fans of 90 kW, and all the CT fans were in service, but during the winter eight to nine are in service. The AP used by the CWP and CT fans is 1.8%–1.9%, which is higher than a CW system with NDCT.

In another typical power plant, CT lift pumps are installed near the CT to lift the circulating water coming from the condenser discharge to provide the necessary head to reach the CW water on top of the CT. Figure 4.8 shows the schematic diagram of a CW system with cooling tower low-pressure (CTLP) pumps. The CTLP pumps are in series with CW pumps (this concept is old and nowadays is not being practiced). There are two CW pumps of 950 kW and two CTLP pumps of 710 kW for a 210 MW power plant. All CW pumps and CTLP pumps are working continuously. The plant is provided with one CT with 36 CT fans of 45 kW, which are in service continuously. The AP used by the CWP and CTLP systems is 1.6%–1.7% of gross generation and the AP used by CT fans is 0.6%–0.7%. The overall AP used by the CWP, CTLP, and CT fans for one 210 MW plant is 2.2%–2.4% of gross generation, which is very high compared with a single CWP system instead of a CWP + CTLP system.

4.4.2 Cooling Tower

The CT fan blades are made of different materials such as aluminum, wood, and glass reinforced plastic (GRP). With the advent of fiber-reinforced plastic (FRP), the CT fan blades are made of FRP [44,45]. These FRP blades are made with good quality epoxy resin to resist the very corrosive air system and a polypax coating with high resistance to ultraviolet degradation and abrasion resistance. The fan achieves a low noise level due to the aerofoil profile of the blades, and use of nonresonant material and a highly polished surface leads to better performance.

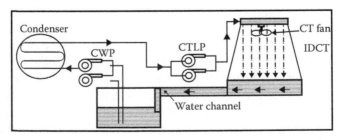

FIGURE 4.8
Schematic diagram of a CW system with cooling tower low-pressure (CTLP) pump.

The CT performance mainly depends on the wet-bulb temperature, air-flow through the CT, and energy consumed by the fan motors. To improve the performance of the CT, initially the fan blades were changed from GRP to FRP blades. The airflow through the fan depends mainly on the pitch angle of the blade, the air gap between the tip of the blade and the casing, and so forth. The pitch of the FRP blade is increased optimally to increase the airflow through the system and optimize the air gap to reduce the air leakage [46,47]. The higher lift to drag ratio, larger chord width along the blade twist, and lower drag losses of FRP fans improve the fan efficiency considerably. The increased fan efficiency and reduced weight of fan blades reduce the electrical power input to the fan motor.

4.4.2.1 Range

The range is the temperature difference between the circulating water inlet and outlet temperature. The cooling load is directly proportional to the range of the CT, which is one of the most important performance indices, and helps in evaluating the performance of the CT. The range of the CT is computed as follows:

$$R = CWT_{in} - CWT_{out} \qquad (4.2)$$

where CWT_{in} is circulating water temperature at the CT inlet in °C and CWT_{out} circulating water temperature at the CT outlet in °C.

The performance of the GRP- and FRP-bladed fans was monitored through-out the year for different weather conditions for a typical 210 MW power plant. Figure 4.9 shows the variation of range for GRP- and FRP-bladed fans. The range changed with the change in wet-bulb temperature and improved from 2.5°C–16.0°C to 4.5°C–17.0°C after fan blade replacement. During the performance test conducted on the CT of Unit 1 (CT1) and Unit 4 (CT4) at a

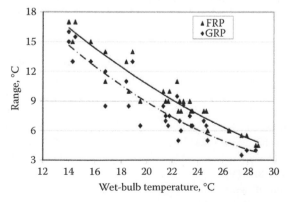

FIGURE 4.9
Variation of range of GRP- and fiber-reinforced plastic (FRP)–bladed fans.

typical 210 MW power plant, the range of CT1 improved from 9.7°C to 11.6°C and CT4 improved from 9.5°C to 11.4°C (see Table 4.9) due to a change to FRP blades. The average overall improvement was about 20% [48].

4.4.2.2 Approach

The approach is computed as the difference between the circulating water temperature at the CT outlet and the wet-bulb temperature in °C:

$$A = CWT_{out} - WBT \tag{4.3}$$

where WBT is the wet-bulb temperature. The approach is also one of the performance parameters for the CT and indicates how close the CW outlet temperature is to the ambient wet-bulb temperature. The main aim of the CT is to keep the approach as low as possible. Figure 4.10 shows the variation of the approach with the variation in wet-bulb temperature. During the performance test the approach of CT1 is reduced from 11.0°C to 9.7°C and CT4 is reduced from 11.0°C to 9.6°C (Table 4.9). The average reduction of the approach was about 12%.

TABLE 4.9

Performance Results of Replacement of GRP Fans by Fiber-Reinforced Plastic (FRP) Fans

		CT1		CT4	
S. No.	Particulars	GRP	FRP	GRP	FRP
1	No. of fans in service	9	9	9	9
2	Average wet-bulb temperature (°C)	23.5	23.8	21.52	22.1
3	Average dry-bulb temperature (°C)	26.9	26.1	31.4	30.8
4	Average CW inlet temperature (°C)	44.2	45.1	42.05	43.1
5	Average CW outlet temperature (°C)	34.5	33.5	32.53	31.73
6	Condenser absolute pressure (kPa)	9.5	9.4	9.3	9.6
7	Condensate hot well temperature (°C)	45.8	46.1	46.5	45.4
8	Total power (kW)	522.63	342.99	487.17	316.35
9	Reduction in power (%)	–	34.37	–	35.06
10	Average air velocity (m/s)	6.99	9.10	8.57	9.57
11	Total air flow (m³/s)	4632.1	6030.4	5679.2	6341.8
12	Increase in air flow (%)	–	30.19	–	11.67
13	Range (°C)	9.7	11.6	9.5	11.4
14	Approach (°C)	11.0	9.7	11.0	9.6
15	Effectiveness (%)	46.86	54.46	46.37	54.14
16	Increase in effectiveness (%)	–	7.6	–	7.77
17	SEC (W/t of air)	26.12	13.17	19.86	11.55
18	Reduction in SEC (%)	–	49.58	–	41.84
19	Average fan efficiency (%)	32.55	64.58	42.82	73.63
20	Increase in fan efficiency (%)	–	32.03	–	30.81

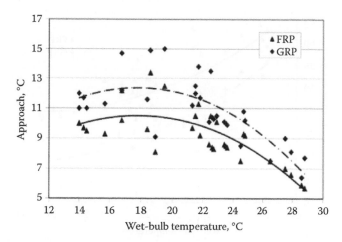

FIGURE 4.10
Variation of approach of GRP- and FRP-bladed fans.

4.4.2.3 Effectiveness

The effectiveness of the CT helps in predicting the thermal efficiency of the CT fans and is computed as follows:

$$EFF = \frac{CWT_{in} - CWT_{out}}{CWT_{in} - WBT} \times 100 = \frac{R}{R+A} \times 100\% \qquad (4.4)$$

Figure 4.11 presents the variation of effectiveness of FRP- and GRP-bladed fan CTs with wet-bulb temperature. The heat transfer effectiveness decreases with an increase in wet-bulb temperature. The effectiveness increased after the change of fan blades from GRP to FRP from 24.5%–59.3% to 39.5%–64.2%. The net improvement of effectiveness is about 15%, which is quite good. The effectiveness of FRP-bladed CT fans is increased from 46.9% to 54.5% for CT1 and improved from 46.4% to 54.1% for CT4 during the performance test (Table 4.9). The effectiveness was improved by about 8%.

4.4.2.4 Specific Energy Consumption

The SEC is an important performance index, as it helps in evaluating the power required by the fans to induce unit airflow through fans, and is computed as follows:

$$SEC = \frac{P \times 10^3}{\overset{o}{m} \times 3600 \times \rho} \text{ kWh/t of air} \qquad (4.5)$$

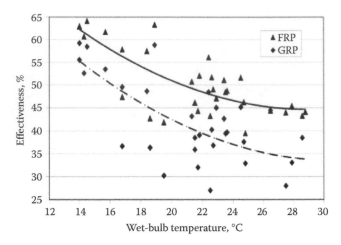

FIGURE 4.11
Variation of effectiveness of GRP- and FRP-bladed fans.

where P is the power in kW, $\overset{o}{m}$ the mass flow of air in m³/s, and ρ the density of air in kg/m³.

Figure 4.12 shows the variation of SEC with the wet-bulb temperature. The SEC increases as the wet-bulb temperature increases. The SEC of FRP fans is reduced considerably because of power reduction as well as increased air-flow in the CT system. The SEC is reduced from 24.3–28.5 kWh/t to 12.64–14.22 kWh/t of air handled. The air throughput is increased by 11.7%–30.2% and the energy consumption is reduced by 34.4%–35.1%. The SEC is reduced by about 40%–50%.

4.4.2.5 Fan Efficiency

The fan efficiency is an important performance index, giving the overall performance of the CT fans, that is, energy consumption and heat removal capacity.

Figure 4.13 shows the variation of fan efficiency with wet-bulb temperature for GRP and FRP fans. It can be seen from the figure that the efficiency decreases as the wet-bulb temperature increases, i.e., it requires more power. By replacing the GRP fan blades with FRP fan blades the efficiency increased from 24.5%–59.3% to 71.7%–80.7%.

4.4.2.6 Performance Results of Replacement of GRP Fan Blades with FRP Fan Blades and Optimum Motor

After replacement of GRP fans by FRP fans, the load factor of the motor was found to be 45.4%–56.3%. Because of the poor loading of the motor, the power factor of the motor is reduced to 0.76–0.80. A detailed study has

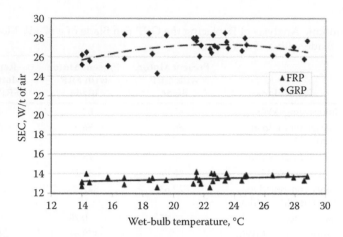

FIGURE 4.12
Variation of SEC of GRP- and FRP-bladed fans.

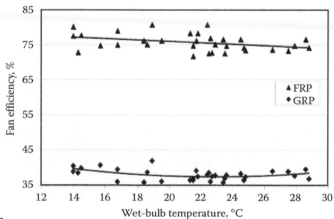

FIGURE 4.13
Variation of efficiency of GRP- and FRP-bladed fans

been carried out to replace the existing 67 kW motor with a 45 kW one [49]. A techno-economic evaluation of the motor replacement along with blade replacement was carried out and is presented in Table 4.10. As a trial, one fan motor is replaced with a 45 kW motor, and the performance is monitored (the performance results are given in Table 4.11). The motor loading increased to 75%–80% and the power factor improved to 0.85–0.91.

The adopted FRP fan blades reduced energy consumption by 129.34 MWh/month. The payback period for an initial investment of US$1587 (9 for

TABLE 4.10

Techno-Economic Analysis of Replacing the GRP Fan Blade of a 67 kW Motor with an FRP Fan Blade with a 45 kW Motor

S. No.	Particulars	Present Motor with GRP Blade	Present Motor with FRP Blade	Replaced Motor with FRP Blade
1	Motor rating (kW)	67	67	45
2	Average motor load factor (%)	86.7	56.9	69.07
3	Motor efficiency (%)	89	75	92
4	Motor input (kW)	58.07	38.11	31.08
5	Energy consumed for nine fans (MWh/day)	522.63	342.99	279.71
6	Power factor	0.85	0.75	0.82
7	Energy savings (MWh/day)	–	179.64	242.92
8	Energy savings (MWh/month)	–	129.34	174.90
9	Savings (US$/month) (energy generation cost at the rate of US $0.029/kWh)	–	3,787.23	5,127.66
10	Investment for FRP fan blades and motors (US$)	–	71,808.51	81,382.98
11	Payback period (months)	–	23	19

one 210 MW plant) was 23 months. The replacement of both fan blades and the motor enhanced the energy savings to 174.9 MWh/month. The capital investment was US$1798 and the payback period was 19 months. The latter is more attractive.

4.5 Water Treatment Plant

The main aim of the WTP in a thermal plant is to provide the demineralized (DM) water for producing steam and treated water for general application service water. The average AP used by the WTP varies from 0.10% to 0.16% of gross generation (see Figure 4.14). The raw water is pumped from the river to the WTP by a number of river water pumps, and the raw water is treated in different stages such as clarifying and demineralization. The AP and the chemical used by the WTP depend on the hardness of the incoming raw water.

In a typical 2 × 210 MW power plant, there are five river water pumps (HT) of 355 kW, of which two pumps work continuously (the river water pump

TABLE 4.11

Comparison of Performance Results of a New 45 kW Fan Motor with an Existing 67 kW Motor

S. No.	Particulars	GRP Fan with Existing 67 kW Motor	FRP Fan with Existing 67 kW Motor	FRP Fan with New 45 kW Motor
1	Average wet-bulb temperature (°C)	23.9	23.9	23.9
2	Average CW inlet temperature (°C)	44.6	44.5	44.5
3	Average CW outlet temperature (°C)	34.8	33.3	33.5
4	Power (kW)	59.8	37.9	31.08
5	Motor loading (%)	89.25	56.57	69.07
6	Power factor	0.87	0.78	0.82
7	Frequency (Hz)	49.9	49.9	49.9
8	Average line voltage (V)	419	419	419
9	Average current (A)	94.7	67.0	52.2
10	Average motor body temperature (°C)	36.5	35.4	37.1
11	Average fan body temperature (°C)	37.8	36.8	37.5
12	Average motor bearing temperature (°C)	39.1	37.8	38.9
13	Average fan bearing temperature (°C)	38.7	38.5	39.1
14	Average air velocity (m/s)	6.85	9.4	9.3
15	Total air flow (m³/s)	504.37	692.13	684.77
16	Range (°C)	9.8	11.2	11.0
17	Approach (°C)	10.9	9.4	9.6
18	Effectiveness (%)	47.34	54.37	53.40
19	SEC (W/t of air)	27.45	12.68	10.51

house is located about 5–6 km away from the power house). Figure 4.15 shows the variation of specific raw water consumption with PLF at a typical power plant. The specific raw water consumption varies between 2.8 and 3.6 L/kWh. As the PLF increases, the specific raw water consumption decreases. The water from the river is pumped to an intermediate (common) one-day reservoir, which is located inside the power house near the WTP. From the one-day reservoir tank, the raw water is pumped to the DM plant by raw water pumps (RWPs) (LT) of 110 kW (two are installed for each unit and one is working continuously). There are five softener feed water pumps of 55 kW to soften the raw water received from the river to feed the DM plant and general purpose service water. Two DM transfer pumps of 55 kW are installed to transfer the DM water from the WTP to an individual unit DM makeup tank. Figure 4.16 shows the variation of specific DM water makeup with PLF at a

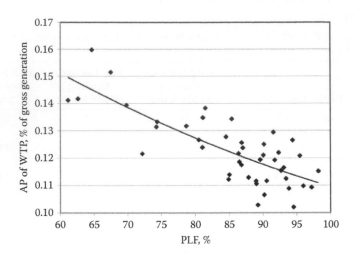

FIGURE 4.14
Variation of AP of water treatment plant (WTP) with PLF.

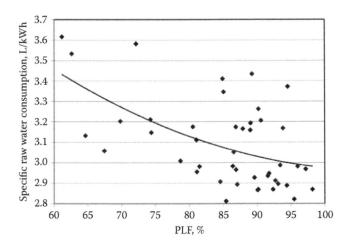

FIGURE 4.15
Variation of specific raw water consumption with PLF.

typical power plant. The specific DM water makeup varies between 0.04 and 0.09 L/kWh. As the PLF increases, the specific DM water makeup decreases. All other auxiliary equipment such as the degasser and regeneration are provided with smaller capacity pump motors.

Tables 4.12–4.15 give the performance results of pump motors at the WTP. The observations from the study are as follows:

1. The average AP used by river water pumps is 0.07% of gross generation. The river water pump 1 combined efficiency is low, about 33.0%, which is lower compared with other pumps as well as the design

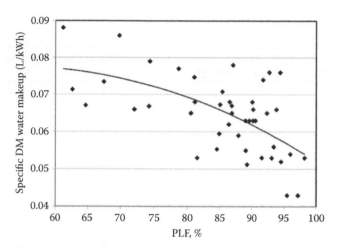

FIGURE 4.16
Variation of specific demineralized (DM) water makeup with PLF.

value of 88.0% due to less water discharge and problems in the pump internals like deformation of impeller and higher clearance between the impeller and casing. *Energy conservation measures*: The replacement of the pump impeller or restoration of pump internals reduces the energy consumption by 60.9 MWh/month. The anticipated investment is US$33,500 and the payback period is 12 months. This energy conservation measure reduces the AP by 0.03% of gross generation.

2. The combined pump efficiencies of river water pumps 2–5 are also on the lower side of 45.1%–55.2% compared with the design value of 88.0% due to higher power input to the pump motors. *Energy conservation measures*: The efficiency of these pumps may also be improved by overhauling of pumps and motors.

3. The average AP used by RWPs is 0.04% of gross generation. There are two RWPs; one is working continuously and the other is kept on standby. The combined efficiency of RWP 2 pump motor is 61.0%, RWP 2 pump motor is 40.1% and are lower side compared to design value of 80.0%. The efficiency of RWP 2 is very low because the discharge pressure developed by the pump is very low (i.e., 73.6 kPa compared with the design value of 215.8 kPa net head), but the discharge flow is on the higher side at about 0.48 m³/s compared with the design value of 0.42 m³/s. *Energy conservation measures*: It is suggested to overhaul RWP 2 or replace it with new pump to get the required discharge pressure. This energy conservation measure reduces the AP by 0.02% of gross generation (41.9 MWh/month). The anticipated investment for a new pump is US$10,500 and the payback period is 6 months.

4. The average AP used by the DM transfer pumps is 0.004% of gross generation. There are two DM transfer pumps, one is working

TABLE 4.12

Performance Results of River Water Pumps

S. No.	Particulars	Unit	Design	Pump 1	Pump 2	Pump 3	Pump 4	Pump 5
01	Pump suction pressure	kPa	–	−92.7	−92.7	−92.7	−92.7	−92.7
02	Pump discharge pressure	kPa	786.8[a]	307.1	317.8	318.8	306.1	309.0
03	River water flow	m³/s	0.33	0.26	0.43	0.39	0.37	0.38
04	Electrical power input	kW	317.2	338.4	321.2	334.9	307.7	338.3
05	Load factor of motor	%	94.0	96.7	91.8	95.7	87.9	96.6
06	Mechanical power	kW	262.3	103.8	177.3	162.3	147.1	152.5
07	Combined efficiency	%	82.6	30.7	55.2	48.4	47.8	45.1
08	SEC	kWh/ m³ of water flow	0.26	0.36	0.21	0.24	0.23	0.25

[a] *Total net head.*

TABLE 4.13

Performance Results of Raw Water Pumps (RWPs)

S. No.	Particulars	Unit	Design	RWP 1	RWP 2
01	Pump suction pressure	kPa	–	−23.5	−23.5
02	Pump discharge pressure	kPa	215.8[a]	174.6	73.6
03	Raw water flow	m³/s	0.42	0.35	0.48
04	Electrical power input	kW	112.5	114.1	116.7
05	Load factor of motor	%	96.0	92.1	94.2
06	Mechanical power	kW	89.9	69.6	46.7
07	Combined efficiency	%	80.0	61.0	40.1
08	SEC	kWh/m³ of water flow	0.08	0.09	0.07

[a] *Total net head.*

TABLE 4.14

Performance Results of Demineralized (DM) Transfer Pumps

S. No.	Particulars	Unit	Design	Pump A
01	Pump suction pressure	kPa	–	–17.9
02	Pump discharge pressure	kPa	392.4[a]	372.8
03	DM water flow	m³/s	135.0	128.8
04	Electrical power input	kW	19.4	21.8
05	Load factor of motor	%	84.9	99.0
06	Mechanical power	kW	14.7	14.0
07	Combined efficiency	%	75.7	64.2
08	SEC	kWh/m³ of water flow	0.14	0.17

[a] *Total net head.*

TABLE 4.15

Performance Results of Softener Water Pumps

S. No.	Particulars	Unit	Design	Pump 1	Pump 2	Pump 3	Pump 4	Pump 5
01	Pump suction pressure	kPa	–	–23.5	–23.5	–23.5	–23.5	–23.5
02	Pump discharge pressure	kPa	245.3[a]	233.5	234.5	238.4	288.4	288.4
03	Water flow	m³/s	0.15	0.14	0.13	0.14	0.13	0.12
04	Electrical power input	kW	47.7	56.9	49.4	56.8	54.6	51.9
05	Load factor of motor	%	81.0	91.9	79.7	91.7	88.1	83.8
06	Mechanical power	kW	37.5	37.1	34.7	35.6	40.4	37.7
07	Combined efficiency	%	78.6	65.3	70.4	62.7	74.0	72.7
08	SEC	kWh/m³ of water flow	0.09	0.11	0.10	0.12	0.12	0.12

[a] *Total net head.*

continuously or intermittently depending on the load on the plant and the other is kept on standby. The water flow at the DM feed water pump is 128.8 m³/h compared with the design value of 135 m³/h and is normal. The combined efficiency of the pump motor is 64.2% and is slightly lower than the design efficiency of 75.7%. The motor loading is also full load, that is, operating without margin, and is higher compared with the design value of 84.9%.

5. The average AP used by the softener water pumps is 0.02% of gross generation. There are five softener water pumps; two are working continuously and the other three are kept on standby. The discharge water flow of the softener feed water pumps is 0.12–0.14 m^3/s compared with the design flow of 0.15 m^3/s. The combined efficiency of these pumps is 62.7%–74.0% and is slightly lower compared with the design value of 78.6%.

4.6 Conclusions

Common AP is the power used to drive the equipment that is common for more than one unit in the entire power station. The running of common auxiliary equipment does not depend on the operation of a particular unit. The power supply to this equipment is fed from STs. The common AP varies between 1.62% of gross generation at MCR condition and 2.29% at 70% PLF. The common AP used by various components is as follows: CHP, about 0.12%–0.22% of gross generation; AHP, about 0.10%–0.25% of gross generation; CWP, about 1.2%–2.4% of gross generation; and WTP, about 0.10%–0.16% of gross generation. Some of the energy conservation measures for common auxiliaries are as follows:

1. The use of a beetle charger or side arm charger reduces the time taken for positioning wagons and also reduces the net energy consumption as well as the demurrage charges of coal rakes from railways.

2. The optimal loading of crushers (above 70%) reduces the SEC and also reduces the AP of the CHP.

3. The use of a bypass screen, that is, vibratory screens, before secondary crushers reduces the energy consumption of crushers by about 20%–30%.

4. Providing and maintaining the screen at the bottom of a secondary crusher to avoid the escape of raw coal bigger than 25 mm from crushers reduces the SEC of mills, avoids mill outages, reduces mill rejects, and so forth.

5. The SEC of conveyor belts is high due to underloading of conveyor belts. Optimum loading of conveyor belts by an automatic monitoring system reduces the SEC of conveyor belts. The load factors of conveyor motors are very low due to underloading of the conveyor system. The poor loading of the motor increases the motor losses by

about 10%–15%. The use of intelligent motor controllers for LT conveyor motors enhances the energy efficiency of the conveyor system.

6. Conveyor belts are placed on top of rollers, and these rollers are not free running and stuck, which causes higher frictional loss. These rollers should be free running to reduce the frictional losses for conveyor belts, which in turn will reduce the energy consumption of belts.

7. The periodic inspection of conveyor belts, adjusting the belts, and clearing the coal on either side of the conveyor belts helps in reduction of energy consumption.

8. The water to ash ratio is 5:1 to 15:1 in many power plants. The increased water to ash ratio increases the pumping power, and the increase in the ratio to 15:1 increases the power by 75% compared withs 5:1 ratio. A ratio that is too low reduces the viscosity of slurry and leads to an increase in frictional loss in the piping system and erosion of the pump impeller. Therefore, the water to ash ratio must be optimized.

9. CW pump impellers can be coated with a surface coating, which reduces pump internal losses, improves pump efficiency, and reduces the SEC of pumps. The anticipated energy savings is 111 MWh/month for a 2 × 210 MW plant.

10. The AP used by a CWP and CTLP system is higher compared with a single CW pump system. The use of a single CWP system reduces the AP by about 0.3%–0.6% gross generation compared with a CWP and CTLP system.

11. The adopted FRP fan blades reduce energy consumption by 129.34 MWh/month. The replacement of both fan blades and the motor enhanced the energy savings to 174.9 MWh/month.

12. The replacement of the pump impeller for river water pumps or restoration of pump internals reduces the energy consumption by 60.9 MWh/month.

The implementation of energy conservation measures for common AP reduces the average AP by 0.4%–0.7% of gross generation.

Part II

Thermal Power Plant Control Process Modeling

5

Physical Laws Applied in Fossil Fuel Power Plant Processes

5.1 Introduction

The combustion process of a coal-fired power plant is highly complex, involving chemical reactions, heat transfer, and slagging. Figure 5.1 shows the coal-fired power plant combustion process.

5.2 Heat Conduction, Convection, and Radiation

Heat transfer including conduction, convection, and radiation occurs in different sections of the furnace. In the center of the furnace, heat transfers to the metal surface of water wall pipes from the flames of pulverized coal burning by radiation. Then the heat transfers to the water side of the metal pipes through conduction and the heat can be absorbed by the flowing water or mixture of steam and water by conduction and convection. In the flue-gas path, heat is carried to the metal surface of the superheater and reheater. Then the heat can be absorbed by convection. Finally, the steam inside the pipes of the superheater or reheater can absorb the heat by conduction and convection. The entire process of pulverized coal combustion is modeled in the research.

Equation 5.1 models the conductive and convective heat transfer [19–21]:

$$\frac{\partial x}{\partial t}(\rho E) + \nabla\left(\overline{v(\rho E + p)}\right) = \nabla \cdot \left(k_{\text{eff}}\nabla T - \Sigma_j h_j \vec{J}_j + \left(\overline{\overline{\tau}}_{\text{eff}} \cdot \vec{v}\right) + S_h\right) \quad (5.1)$$

where ρ is the intensity, p the pressure, $\vec{v}$ the velocity, k_{eff} the effective conductivity, $\vec{J}_j$ the diffusion flux of species j, h_j the enthalpy of species of j, $\overline{\overline{\tau}}_{\text{eff}}$ the viscosity of the flue, and term S_h the amount of heat from the chemical reaction and any other heat source defined. In Equation 5.1, the energy transfer due to conduction is defined as

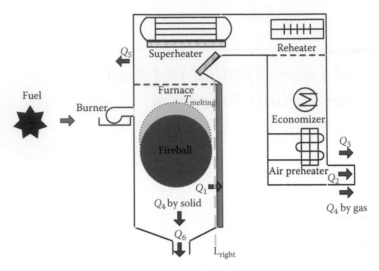

FIGURE 5.1
Combustion process and slagging in the furnace of a coal-fired power plant boiler.

$$q_{cond} = k_{eff} \nabla T \tag{5.2}$$

The energy transfer due to species diffusion is defined as

$$q_{diff} = \Sigma_j h_j \vec{J}_j \tag{5.3}$$

The energy transfer due to viscous dissipation is defined as

$$q_{diss} = \bar{\tau}_{eff} \cdot \vec{v} \tag{5.4}$$

In Equation 5.1,

$$E = h - \frac{p}{\rho} + \frac{v^2}{2} \tag{5.5}$$

where sensible enthalpy h is defined for an ideal gas as

$$h = \Sigma_j Y_j h_j \tag{5.6}$$

In Equation 5.6, Y_j is the mass fraction of species j and

$$h_j = \int_{T_{ref}}^{T} c_{p,j} \, dT \tag{5.7}$$

where T_{ref} is 298.15 K and $c_{p,j}$ is the specific heat capacity rate of species j.

Pulverized coal combustion is a nonadiabatic, non-premixed process, and the total enthalpy form of the energy in the model is given as

$$\frac{\partial y}{\partial \chi}(\rho H) + \nabla \cdot (\rho \vec{v} H) = \nabla \cdot \left(\frac{k_t}{c_p} \nabla H \right) + S_h \tag{5.8}$$

where ρ is the intensity, p is the pressure, $\vec{v}$ is the velocity, k_t is the conductivity of flue gas in turbulent combustion, and c_p is the specific heat capacity rate. In Equation 5.8, the total enthalpy H is defined as

$$H = \Sigma_j \, Y_j \, H_j \tag{5.9}$$

where Y_j is the mass fraction of species j and

$$H_j = \int_{T_{ref\,j}}^{T} c_{p,j} \, dT + h_j^0 \left(T_{ref,j} \right) \tag{5.10}$$

where $h_j^0 \left(T_{ref,j} \right)$ is the formation enthalpy of species j at the reference temperature T_{ref}.

In Equation 5.8, the chemical reaction energy source S_h is defined as

$$S_h = -\Sigma_j \frac{h_j^0}{M_j} R_j \tag{5.11}$$

where h_j^0 is the enthalpy of formation of species j, R_j is the volumetric rate of creation of species j, and M_j is the molecular mass of species j. In the metal pipes of the water wall, superheater, and reheater, the energy equation is given as [21]

$$\frac{\partial}{\partial t}(\rho h) + \nabla \cdot \left(\vec{v} \rho h \right) = \nabla \cdot (k \nabla T) + S_h \tag{5.12}$$

where ρ is the density, h the enthalpy, k the conductivity, T the temperature, and S_h the volumetric heat source.

The radiation transfer equation for an absorbing, emitting, and scattering medium at position $\vec{r}$ in the direction $\vec{s}$ is given as [21–24]

$$\frac{d1(\vec{r},\vec{s})}{ds} + (a + \sigma_s) I(\vec{r},\vec{s}) = an^2 \frac{\sigma T^4}{\pi} + \frac{\sigma_s}{4\pi} \int_0^{4\pi} I(\vec{r},\vec{s}) \, \varnothing \, (\vec{s} \cdot \vec{s}') \, d\Omega' \tag{5.13}$$

where $\vec{r}$ is the position vector, $\vec{s}$ the direction vector, $\vec{s}'$ the scattering direction vector, s the path length, a the absorption coefficient, n the refractive index, σ_s the scattering coefficient, σ the Stefan–Boltzmann constant

$(5.66 \times 10^{-8} \text{ W} \cdot \text{m}^{-2} \cdot \text{K}^{-4})$, I the radiation intensity, which depends on position $\vec{r}$ and direction $\vec{s}$, T the local temperature, $\varnothing$ the phase function, and Ω' the solid angle.

Energy coupling and the discrete ordinates (DO) model [20] are applied in the research to simulate the heat radiation process inside the furnace. The DO model considers Equation 5.13 in the direction $\vec{s}$ as a field equation. Thus the equation is written as

$$\nabla \cdot (I(\vec{r},\vec{s})\vec{s}) + (a + \sigma_s)I(\vec{r},\vec{s}) = an^2 \frac{\sigma T^4}{\pi} + \frac{\sigma_s}{4\pi} \int_0^{4\pi} I(\vec{r},\vec{s}) \varnothing (\vec{s} \cdot \vec{s}') \, d\Omega' \quad (5.14)$$

The energy equation when integrated over a control volume i can get the model of coupling between energy [21–23,65]. The model is presented as follows:

$$\sum_{j=1}^{N} \mu_{ij}^T T_j - \beta_i^T T_i = \alpha_i^T \sum_{k=1}^{L} I_i^k \omega_k - S_i^T + S_i^h \quad (5.15)$$

where $\alpha_i^T = k\Delta V^i$, $\beta_i^T = 16k\sigma T_i^3 \Delta V_i$, $S_i^T = 12k\sigma T_i^4 \Delta V_i$, k the absorption coefficient, and ΔV the control volume. The coefficient μ_{ij}^T and the source term S_i^h are due to the discretization of the convection and diffusion terms.

The research focuses on optimizing coal-fired combustion process in which pulverized coal and oxide air enter the reaction zone in distinct streams. Compared with a premixed system in which reactants are mixed at the molecular level before reaction, pulverized coal combustion is a nonpremixed system, so a nonpremixed combustion model [18–22] is applied in the research. The basis of the model is that the instantaneous thermochemical state of the fluid is related to a conserved scalar quantity known as the mixture fraction, f, which is given as

$$f = \frac{z_i - z_{i\,\text{ox}}}{z_{i\,\text{fuel}} - z_{i\,\text{ox}}} \quad (5.16)$$

where z_i is the element mass fraction for element i. The subscript ox denotes the value at the oxidizer stream inlet and the subscript fuel denotes the value at the fuel stream inlet. The transport equation for the mixture fraction [21–23], [50] is given as

$$\frac{\partial}{\partial t}\left(\rho \overline{f}\right) + \nabla \cdot \left(\rho \vec{v} \overline{f}\right) = \nabla \cdot \left(\frac{\mu_t}{\sigma_t} \nabla \overline{f}\right) + S_m \quad (5.17)$$

where ρ is the density, $\overline{f}$ is the mean mixture fraction, $\vec{v}$ is the local velocity, μ_t is the turbulent viscosity, the constant $\sigma_t = 0.85$, and the source term S_m is solely due to transfer of mass into the gas phase from the pulverized coal particle.

5.3 Heat Balance

How and where does the heat get lost and emissions increase inside the boiler? Equation 5.18 gives an expression of the heat balance in the combustion process of boilers [18]:

$$Q_1 + Q_2 + Q_3 + Q_4 + Q_5 + Q_6 = 100\% \tag{5.18}$$

Q_1 is the heat absorbed by the water and steam inside the pipes of the water wall, superheater, reheater, and economizer. It also includes the heat recovered from the preheater where the cold air absorbs the heat of the residual flue gas. Normally, Q_1 is 75%–90%. However, slagging that accumulates on the heat transfer surface can severely influence Q_1. More slagging on the surface of the water wall, superheater, and other heat transfer equipment can massively decrease the heat radiation by which the heat radiates from the flame of the fireball to the heat transfer surface. At the same time, more slagging can cause less heat conductivity in which the heat is not more massively and rapidly transferred to the water or steam side than a system with less slagging on the surface of the heat transfer equipment.

In addition, more slagging can increase the blockage in the flue-gas pass and decrease the convection in which heat can rapidly convect to other heat transfer equipment from the surface of the fireball. Moreover, more blockage caused by the slagging in the flue-gas pass can increase the power consumption of the forced draft (FD) and induced draft (ID) fans, which decreases the overall efficiency of the power plant. The proposed solution in this research tries to solve the problem by limiting slagging formation. Figure 5.2 shows the slagging deposition bonds on the surface of the water wall in a boiler furnace.

FIGURE 5.2
Slagging accumulation on the water wall surface of a boiler furnace.

Figure 5.3 shows fouling accumulated on the surface of the superheater and reheater in the convection pass area of a boiler furnace.

Slagging is built up on furnace walls, which are mainly in the radiation section. It is in a highly viscous state and forms a liquid layer. Fouling is built up by condensed materials. It is a dry deposit and generally in the convection section.

Q_2 is the heat carried by the exit flue gas, which includes the water vapor, oxygen, nitrogen, carbon dioxide, and other gas transporting the residual heat to the atmosphere. A higher temperature of the exit gas and a larger exit gas volume can cause more residual heat loss from the furnace of a boiler. On the other hand, a too low temperature of exit gas can cause more chemical corrosion to the surface of heat transfer equipment installed in the flue-gas pass near the exit. The two outputs of heat loss and corrosion conflict with each other with some of the same parameters such as exit gas temperature and pressure. The proposed solution in this research tries to maintain an exact temperature and volume of exit gas by controlling the position of the fireball, the speed of the flue gas, and the excess air rate.

Q_3 is the heat contained in the combustible gases like CO, H_2, and CH_4 that are unburned and emit with the exit flue gas. The proposed solution in this research tries to adjust the position and temperature of the fireball by

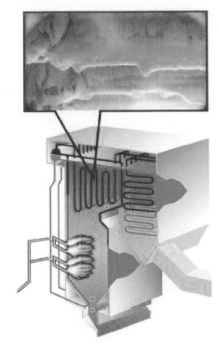

FIGURE 5.3
Fouling deposits in the superheater and reheater area, which are the hottest parts of the convection pass area.

controlling the speed, temperature, and amount of the mixture of primary air and coal powder, and the speed, amount, and pressure of secondary air to maintain a complete burning.

Q_4 is the heat contained in the carbon, which is unburned and lost with clinkers dropped outside of the furnace. The proposed solution in this work tries to maintain the powder cloud of the mixture of coal and air a little longer in the exact position of the furnace and keep an exact temperature of the fireball by controlling the speed of the rotating fireball, the amount of fuel flow in the pipe, and the angle of the burners.

Q_5 is the heat loss from outside of the furnace of the boiler. Normally, it can be decreased by improving the insulation conditions outside the surface of the furnace and it is much less than Q_2, Q_3, and Q_4, respectively.

Q_6 is the heat carried by the clinkers, which are dropped outside of the furnace from the furnace bottom; this is much less than Q_2, Q_3, and Q_4, respectively.

Therefore, this research will aim to effectively minimize heat losses Q_2, Q_3, and Q_4, respectively, by controlling the related input parameters to maintain an optimum fireball. In addition, the research will provide effective methods to maximize heat Q_1 by maintaining maximum fireball flame heat radiation and minimizing slagging and fouling accumulation. Boiler combustion efficiency will be increased from the two areas of improvement.

5.4 Mass Balance

The differential equations of a coal-fired power plant model have been developed [51]. The model is simplified by assuming that time derivatives of variables are considered while spatial derivatives are not. The superheated steam and furnace exhaust gasses are treated as ideal gases. The model is supported by the basic physical thermal dynamics balances as follows.

Heat balance for the superheater, reheater, tubes of the water wall, and economizer in the model is given as

$$Q_{in} + w_{in} h_{in} = w_{ou} h_{ou} + V \frac{d}{dt}(\rho h_{ou})$$ (5.19)

where Q_{in} is the incoming heat (J/s), w_{in} the inlet mass flow (kg/s), h_{in} the inlet-specific enthalpy (J/kg), w_{ou} the outlet mass flow (kg/s), h_{ou} the outlet-specific enthalpy (J/kg), ρ the specific density (kg/m^3), and V the volume (m^3). Mass balance in the model is given as

$$w_{in} - w_{ou} = \frac{d}{dt}(\rho V)$$ (5.20)

The variable definitions in the equation are the same as these of Equation 5.19. The equation of heat radiation in the model is from the Stefan–Boltzmann law:

$$Q = K\theta w_g T_g^4 \frac{1}{\rho_g} \tag{5.21}$$

where Q is heat flow from combustion flame radiation (J/s), K is the coefficient ($K = 0.18$), θ is the specific angle (rad), w_g is the flow of substances entering combustion (kg/s), T_g is the temperature of the flue gas (K), and ρ_g is the density of the combustion flue gas (kg/m³).

The equations of heat transfer due to convection in the model are from engineering experiment. The equation for the heat transfer from combustion gas to the surface of metal tubes is given as

$$Q = K w_g^{0.6} (T_g - T_m) \tag{5.22}$$

where T_m is the temperature of the surface of metal tubes of the heat exchanger such as superheaters and reheaters (K). The definitions of other variables are the same as these of Equation 5.21. The combustion flue gas is assumed as turbulent gas flow.

The equation for the heat transfer from the surface of metal tubes to the steam is given as

$$Q = K w_s^{0.8} (T_m - T_s) \tag{5.23}$$

where T_s is the temperature of steam (K), and the definitions of other variables are the same as those of Equation 5.22. The steam is assumed as turbulent steam flow in the model.

Although the model showed satisfactory results, it cannot be used to solve slagging-related boiler combustion problems because slagging is not considered in the model. Furthermore, the spatial derivatives are also not considered in the model. For example, the model cannot be used to identify the slagging distribution in the furnace of the boiler because slagging is spatially distributed on the heat transfer surface of the water walls of a boiler.

5.5 Turbulent Combustion Gas Flow and Steam Flow

Slagging occurs in the radiant section of a coal boiler with a high temperature, and it is usually associated with some degree of melting of the ash [52]. In coal-fired power plant boilers, slagging can occur on the furnace water walls and the first few rows of superheater tubes. The aerodynamics of the flue gas in the combustion process can convey ash particles to the vicinity of the heat transfer, and the ash particles can pass to the boundary area by inertia. Figure 5.1 shows the boundary area L_{right} that is close to the right side of the furnace. The ash particles can adhere to the surface of water wall tubes if either the particles or the surface is "sticky" enough to overcome the kinetic energy of the incoming particles, and prevent it from rebounding from the heat transfer surface [52]. Therefore, maintaining an appropriate temperature in the boundary of the furnace and keeping the incoming particles from melting can decrease slagging.

Based on these mechanisms of heat transfer, chemical reactions, and slagging, the research proposes a novel way to improve coal-fired power plant boiler efficiency and decrease slagging.

5.6 Conclusion

The main physical laws applied to thermal power plant modeling, control, and improving energy efficiency are discussed in this chapter. Further detailed methods for power plant process modeling, simulation, control, and efficiency improvement are discussed in Chapters 6 through 10. Many common modeling tools including MATLAB®, MATLAB Simulink®, VisSim, Comsol, ANSYS, and ANSYS Fluent are used to process the data of thermal power plants.

6

Modeling and Simulation for Subsystems of a Fossil Fuel Power Plant

6.1 Introduction

The complex power plant system involves many subsystems such as the fuel-handling system, water-handling system, boiler-combustion system, boiler-combustion-security system, ash-handling system, dust-handling system, steam-generation system, turbine system, and generator system. This project focuses on four subsytems: the boiler-combustion-optimization model, boiler control, steam-temperature control, and boiler-turbine-generator model. Physical principles including energy conservation, momentum conservation, and mass balance are applied to develop the models. In addition, the genetic algorithm (GA) integrated with computational fluid dynamics (CFDs) is used to create a boiler-combustion-optimization model.

MATLAB/Simulink and CFD analysis software Comsol are widely used in both industry analysis and scientific research. This software is integrated to develop a boiler-combustion-optimization model in this project. MATLAB/Simulink is also used to develop boiler-control models. VisSim software is used to develop steam-temperature control and an integrated boiler-turbine-generator model in this project.

6.2 Development of a Boiler System Model

Figure 6.1 shows the block diagram of a boiler with a furnace, riser, drum, superheater and attemparator, and reheater with input and output parameters and the interconnected process loops of the boiler system [51,53]. This block diagram has been broadly used to develop a boiler-control model using MATLAB/Simulink software [53].

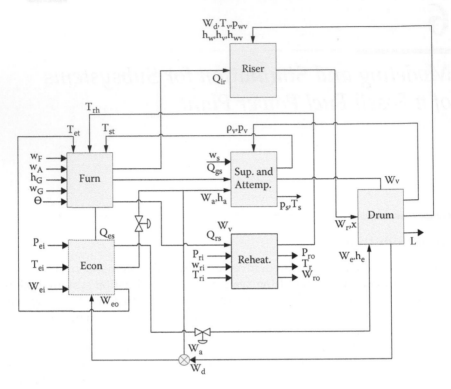

FIGURE 6.1
Complete block diagram of a boiler.

The physical principles including energy conversation, heat balance, and mass balance of all the subsystems are given [54]. Based on the physical principles, the components of the boiler models are given [55]. Based on the mathematical equations, models are developed using MATLAB/Simulink to simulate the boiler. Figure 6.2 shows the integrated modes of a boiler including the furnace, riser, superheater and attemperator, reheater, and drum.

The details of inputs, parameters, states, and outputs for each component are described in the following.

6.2.1 Furnace Modeling

The furnace, also known as the combustion chamber, releases heat to become the heat-transfer system. Time, temperature, and turbulence are the three main parameters needed for combustion to take place in the furnace. A negative pressure is required to be maintained in the furnace in a balanced draft boiler by controlling the furnace draft [56].

Figure 6.3 shows the developed model of the furnace; the governing equations, inputs, parameters, states, and outputs are listed in the following [51]:

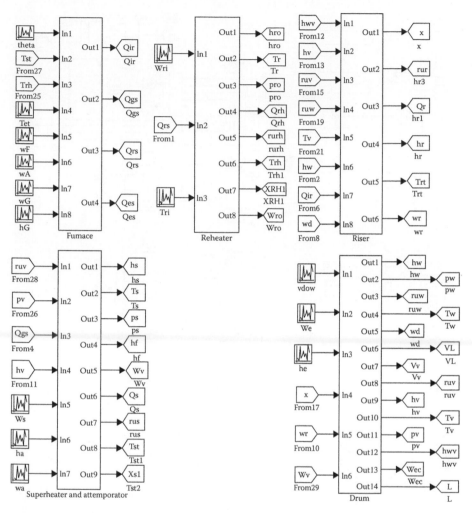

FIGURE 6.2
Integrated model of a power plant boiler system using MATLAB/Simulink.

Equations:
Algebraic Equations:

$$h_{EG} = \frac{x_{F1}}{\rho_{EG}}$$

$$T_g = \frac{h_{EG} - h_{ref}}{c_{pg}} + T_{ref}$$

$$w_{EG} = k_F p_G$$

$$Q_{ir} = \theta k V_F \sigma T_g^4$$

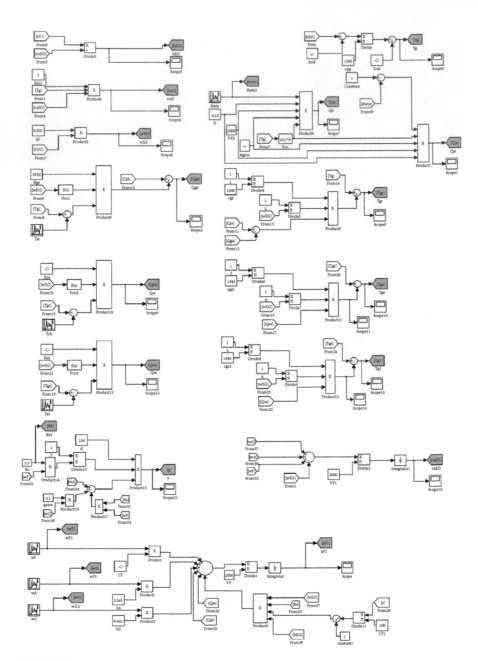

FIGURE 6.3
Model of furnace developed using MATLAB/Simulink.

$$Q_{is} = (1 - \theta)kV_F\sigma T_g^4$$

$$Q_{gs} = Q_{is} + k_{gs}w_{EG}^4\left(T_g - T_{st}\right)$$

$$T_{gr} = T_g + \frac{1}{c_{gs}}\frac{1}{w_{EG}}\left(Q_{is} - Q_{gs}\right)$$

$$Q_{rs} = k_{rs}w_{EG}^{0.6}\left(T_{gr} - T_{rh}\right)$$

$$T_{ge} = T_{gr} - \frac{1}{c_{gs}}\frac{1}{w_{EG}}Q_{rs}$$

$$Q_{es} = k_{es}w_{EG}^{0.6}\left(T_{ge} - T_{et}\right)$$

$$T_{g1} = T_{ge} - \frac{1}{c_{gs}}\frac{1}{w_{EG}}Q_{es}$$

$$y = 100\left(w_A + \gamma w_G - w_F R_s\right)\frac{1}{w_F R_s}$$

Differential Equations:

$$\frac{d}{dt}x_{F1} = \frac{1}{V_F}\left[C_F w_F + h_A w_A + h_G w_G - Q_{ir} - Q_{is} - w_{EG}R_s\left(1 + \frac{y}{100}\right)h_{EG}\right]$$

$$\frac{d}{dt}\rho_{EG} = \frac{1}{V_F}\left(w_F + w_A + w_G - w_{EG}\right)$$

Inputs:

w_F fuel flow to the furnace [kg/s]

w_A air flow to the surface [kg/s]

h_G enthalpy of exhaust gas from gas turbine [J/kg]

w_G exhaust gas flow from gas turbine [kg/s]

θ tilt angle coefficient [$0 < \theta < 1$] rad

T_{st} temperature of superheater metal cubes [K]

T_{rh} temperature of reheater metal cubes [K]

T_{et} temperature of economizer metal tube [K]

h_A inlet air enthalpy [J/kg]

Parameters:

k attenuation coefficient

k_f friction coefficient [m·s]

K_{gs} experimental heat-transfer coefficient to the superheater [J/(kg·K)]
C_{gs} combustion gas specific heat capacity [J·s/(kg·K)]
k_{rs} experimental heat-transfer coefficient to reheater [J/(kg·K)]
V_F combustion chamber volume [m³]
C_F fuel calorific value [J/kg]
R_s stoichiometric air/fuel ratio
γ content of fresh air in exhaust from gas turbine
k_{es} experimental heat-transfer coefficient to the economizer [J/(kg·K)]

States:

x_{F1} $h_{EG} \times \rho_{EG}$ [J/m³]
ρ_{EG} density of exhaust gas from the boiler [kg/m³]

Outputs:

Q_{ir} heat transferred to the risers [J/s]
Q_{is} heat transferred by radiation to the superheater [J/s]
Q_{rs} heat transferred to the reheater [J/s]
Q_{es} heat transferred to the economizer [J/s]
p_G furnace air pressure [Pa]
Q_{gs} total heat transferred to the superheater [J/s]
h_{EG} enthalpy of exhaust gas from the boiler [J/s]
w_{EG} mass flow of exhaust gas from the boiler [kg/s]
T_g gas temperature at the superheater [K]
T_{g1} boiler exhaust gas temperature [K]
y percentage of excess air [%]

6.2.2 Riser Modeling

Figure 6.4 shows the developed model of the riser; the governing equations, inputs, parameters, states, and outputs are listed in the following [51]:

Equations:
Algebraic Equations:

$$x = \frac{h_f - h_{wv}}{h_v - h_{wv}}$$

$$\rho_r = \left[\frac{x}{\rho_v} + \frac{(1-x)}{\rho_{wv}} \right]^{-1}$$

$$Q_r = k_r \left(T_{rt} - T_v \right)^3$$

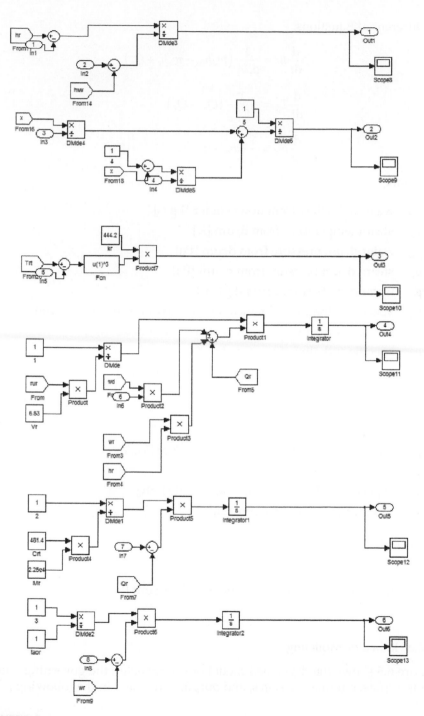

FIGURE 6.4
Model of riser developed using MATLAB/Simulink.

Differential Equations:

$$\frac{d}{dt}hr = \frac{1}{\rho_r V_r}(w_d h_w - w_r h_r + Q_r)$$

$$\frac{d}{dt}T_{rt} = \frac{1}{M_r C_{rt}}(Q_{ir} - Q_r)$$

$$\frac{d}{dt}w_r = \frac{1}{\tau_r}(w_d - w_r)$$

Inputs:

w_d water mass flow from downcomer [kg/s]
T_v steam temperature from drum [K]
p_w water drum pressure from drum [Pa]
p_v steam drum pressure from drum [Pa]
ρ_w water density from drum [kg/m³]
h_w specific enthalpy of downcomer and drum water from drum [J/s]
h_v specific enthalpy of saturated steam from drum [J/s]
h_{wv} specific enthalpy of saturated water from drum [J/s]
Q_{ir} heat transferred to the risers from furnace [J/s]

Parameters:

V_r riser volume [m³]
ρ_r liquid vapor mixture density at the riser [kg/m³]
T_{rt} metal tube temperature [K]
k_r experimental heat-transfer coefficient [J/sK]
τ_r an empirical flow time constant

Outputs:

x steam quality to the drum
w_r liquid vapor mixture mass flow to the drum [kg/s]

6.2.3 Reheater Modeling

Figure 6.5 shows the developed model of the reheater; the governing equations, inputs, parameters, states, and outputs are listed in the following [51]:

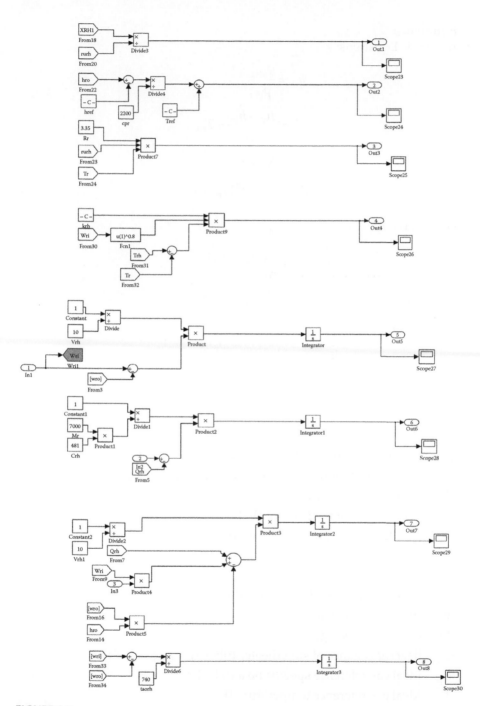

FIGURE 6.5
Model of reheater developed using MATLAB/Simulink.

Equations:
Algebraic Equations:

$$h_{ro} = \frac{x_{RH1}}{\rho_{rh}}$$

$$T_r = \frac{h_{ro} - h_{ref}}{c_{pr}} + T_{ref}$$

$$p_{ro} = R_r \rho_{rh} T_r$$

$$Q_{rh} = K_{rh} w_{ri}^{0.8} (T_{rh} - T_r)$$

Differential Equations:

$$\frac{d}{dt} \rho_{rh} = \frac{1}{V_m} (w_{ri} - w_{ro})$$

$$\frac{d}{dt} T_{rh} = \frac{1}{M_r C_{rh}} (Q_{rs} - Q_{rh})$$

$$\frac{d}{dt} x_{RH1} = \frac{1}{V_{rh}} (Q_{rh} + w_{ri} h_{ri} - w_{ro} h_{ro})$$

$$\frac{d}{dt} w_{ro} = \frac{(w_{ri} - w_{ro})}{\tau_{rh}}$$

Inputs:

p_{ri} pressure of steam at the inlet to the reheater [Pa]
w_{ri} flow of steam at the inlet steam to the reheater [kg/s]
T_{ri} inlet steam pressure [K]
Q_{rs} heat flow from the furnace [J/s]
h_{ri} specific enthalpy of inlet steam [J/kg]

Parameters:

k_{rh} experimental heat-transfer coefficient [J/kg·K]
V_{rh} reheater volume [m³]
M_r reheater mass [kg]
C_{rh} heat capacitance of superheater tubes [J/kg·K]
C_{pr} ideal gas reference specific heat [J/kg·K]
T_{ref} ideal gas reference temperature [K]
h_{ref} ideal gas reference specific enthalpy [J/kg]

States:

ρ_{rh} steam density in the reheater [kg/m³]

T_{rh} reheater metal tube temperature [K]

xRH1 $h_{ro} * \rho_{rh}$

w_{ro} outlet steam mass flow [kg/s]

Outputs:

T_{rh} reheater metal tube temperature [K]

P_{ro} outlet steam pressure [Pa]

T_r reheater steam temperature [K]

H_{ro} specific enthalpy of outlet steam [J/kg]

Q_{rh} heat transferred to the steam [J/s]

ρ_{rh} steam density in the reheater [kg/m³]

x_RH1 $h_{ro} * \rho_{rh}$

w_{ro} flow of steam at the outlet from the reheater [kg/s]

6.2.4 Superheater and Attemperator Modeling

Figure 6.6 shows the developed model of the superheater and attemperator; the governing equations, inputs, parameters, states, and outputs are listed in the following [51]:

Equations:
Algebraic Equations:

$$h_s = \frac{x_{s1}}{\rho_s}$$

$$T_s = \frac{h_s - h_{ref}}{c_{ps}} + T_{ref}$$

$$p_s = R_s \rho_s T_s$$

$$w_v = \sqrt{\frac{(p_v - p_s)\rho_v}{f_s}}$$

$$Q_s = K_s w_v^{0.8} (T_{st} - T_s)$$

Differential Equations:

$$\frac{d}{dt}\rho_s = \frac{1}{V_s}(w_v - w_s)$$

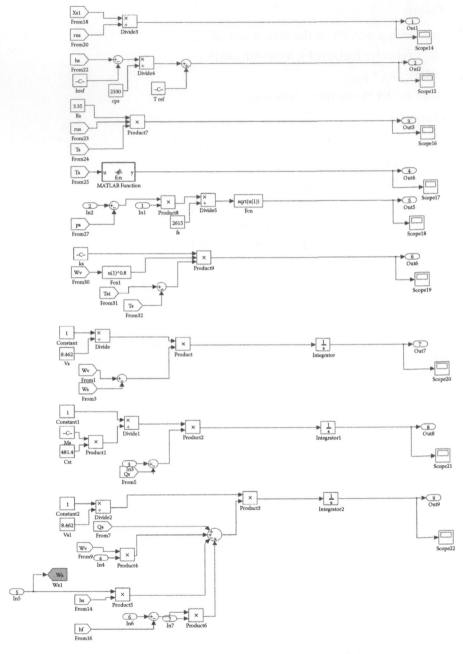

FIGURE 6.6
Model of superheater and attemperator developed using MATLAB/Simulink.

$$\frac{d}{dt}T_{st} = \frac{1}{M_{st}C_{st}}\left(Q_{gs} - Q_s\right)$$

$$\frac{d}{dt}x_{s1} = \frac{1}{V_s}\left[Q_s + w_v h_v - w_s h_s + \left(h_a - h_f\right)w_a\right]$$

Inputs:

W_a attemperation water flow [kg/s]

W_s steam flow from the superheater [kg/s]

p_v steam drum pressure [Pa]

ρ_v density of saturated steam from the drum [kg/m³]

Q_{gs} heat flow from the surface [J/s]

h_v specific enthalpy of saturated steam from the drum [J/kg]

h_a specific heat enthalpy of attemperation water [J/kg]

Parameters:

f_s superheater friction coefficient [m⁻⁴]

k_s experimental heat-transfer coefficient [J/kg·K]

V_s superheater volume [m³]

M_s superheater mass [kg]

C_{st} heat capacitance of superheater tubes [J/kg·K]

C_{pr} ideal gas reference specific heat [J/kg·K]

T_{ref} ideal gas reference temperature [K]

h_{ref} ideal gas reference specific enthalpy [J/kg]

States:

ρ_s density of superheated steam [kg/m³]

T_{st} superheater metal tube temperature [K]

x_{s1} $h_s * \rho_s$

Outputs:

w_v drum outlet steam pressure [Pa]

T_{st} superheater metal tube temperature [K]

p_s pressure of superheated steam [Pa]

T_s temperature of superheated system [K]

h_f specific enthalpy of evaporation [J/kg]

Q_s heat transferred to the steam [J/s]

6.2.5 Drum Modeling

Figure 6.7 shows the developed model of the drum; the governing equations, inputs, parameters, states, and outputs are listed in the following [51]:

Equations:
Algebraic Equations:

$$h_w = \frac{x_{D1}}{m_{dL}}$$

$$p_w = 2.3815 \times 10^6 - 10.1102 h_w + 1.0905 \times 10^{-5} h_w^2$$

$$\rho_w = 1003.4 - 0.58372 \times 10^{-4} h_w - 1.1966 \times 10^{-10} h_w^2$$

$$T_w = 268.3632 + 0.26922 \times 10^{-3} h_w + 0.34182 \times 10^{-6} h_w^2$$

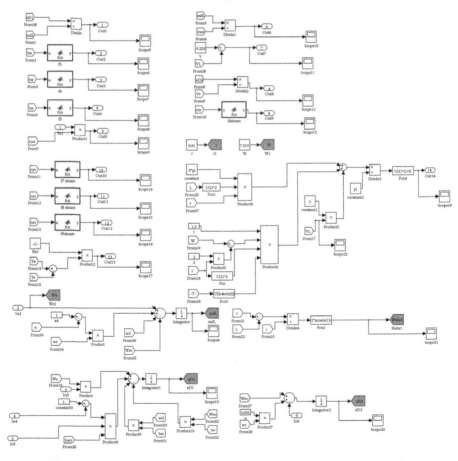

FIGURE 6.7
Model of drum developed using MATLAB/Simulink.

$$w_d = v_{dow} \rho_w$$

$$V_L = \frac{m_{dL}}{\rho_w}$$

$$V_v = V - V_L$$

$$\rho_v = \frac{x_{D2}}{V_v}$$

$$h_v = 268.3632 + 0.26922 \times 10^{-3} \rho_v + 0.34182 \times 10^{-10} \rho_v^2$$

$$T_v = 390.4075 + 35.5266 n_v^2 + 2.7876 n_v^2$$

$$n_v = \ln(\rho_v)$$

$$p_v = 42819 + 217030 \rho_v - 703.6933 \rho_v^2 - 526.3118 \rho_v^3 + 0.7483 \rho_v^4$$

$$h_{wv} = 483140 + 141310 n_v + 16447 n_v^2 + 1373.1 n_v^3$$

$$w_{ec} = k_{ec}(T_w - T_v)$$

$$L = f_{10}^{-1}(V_L) = f_{10}^{-1}\left(\frac{m_{dL}}{\rho_w}\right)$$

$$f_{10}(L) = \frac{1}{3}\pi L^2(3r - L) + \frac{1}{2}(W - 2r)r^2(\theta - \sin\theta)$$

$$\theta = 2\cos^{-1}\left(\frac{r - L}{r}\right)$$

Differential Equations:

$$\frac{d}{dt}m_{dL} = w_e + (1 - x)w_r - w_d - w_{ec}$$

$$\frac{d}{dt}x_{D1} = w_e h_e + (1 - x)w_r h_{wv} - w_d h_w - w_{ec} h_v$$

$$\frac{d}{dt}x_{D2} = w_{ec} + x w_r - w_v$$

Inputs:

h_e specific enthalpy of water from the economizer [J/kg]

V_{dow} volumetric water flow to the downcomer [m³/s]

w_e water flow from the economizer [kg/s]

Q_{ir} heat flow from the furnace [J/s]

w_v steam flow to the superheater [kg/s]

Parameters:

V volume of the drum [m³]

k_{ec} evaporation coefficient [kg/Ks]

R drum radius

w_{ec} steady state evaporation constant [kg/s]

k_r experimental heat-transfer coefficient [J/s·K]

V_r riser volume [m³]

M_r mass of riser metal tubes [kg]

C_{rt} metal specific heat [J/kg·K]

τ_r mass flow time constant [s]

States:

m_{dL} drum liquid mass [kg]

x_{D1} $h_w \times m_{dL}$

x_{D2} $\rho_v \times V_v$

V_v volume of steam in the drum [m³]

h_r liquid vapor mixture specific enthalpy [J/kg]

T_{rt} riser metal tube temperature [K]

w_r liquid–vapor mixture mass flow from the risers [kg/s]

Outputs:

p_v drum outlet steam pressure [Pa]

ρ_v drum outlet steam density [kg/m³]

h_v drum outlet steam specific enthalpy [J/kg]

h_r liquid vapor mixture specific enthalpy [J/kg]

T_{rt} riser metal tube temperature [K]

w_r liquid vapor mixture mass flow from the risers [kg/s]

ρ_w drum water density [kg/m³]

w_d water mass flow to the riser [kg/s]

m_{dL} liquid drum mass [kg]

L drum water level [m]

x steam quality

T_w drum water temperature [K]

6.3 Development of a Boiler System Model Using Simulink

Figure 6.8 [51,53] shows the basic control logic for a boiler.

In the modern power plant, a greater percentage of the control actions that are needed to operate the process are automated. There are several advantages of automating the plant, namely, the reduction of human error in plant operation to provide greater safety for personnel, the reduction of the number of operators needed to operate the plant to reduce labor costs, and lastly, automatic controllers respond more accurately than human operators and respond more quickly to changes in operating conditions. There are two types of control functions used in the plant: the on–off control and the modulating control. On–off control is known as digital control, discrete control, or sequential control. Modulating control is called analog control, continuous control, or closed-loop control. Both these control types complement each other and are essential in the proper operation of the plant. This report will focus on modulating control applications.

The most essential applications of modulating control in the power plant are in the area of boiler control. The aim is to regulate the input of fuel to the boiler, responding to the load demand changes while maintaining essential variables such as steam pressure, steam temperature, and drum water level within acceptable limits. Other areas of application include controlling pressure, temperature, level, and flow variables in the turbine cycle and plant auxiliary systems.

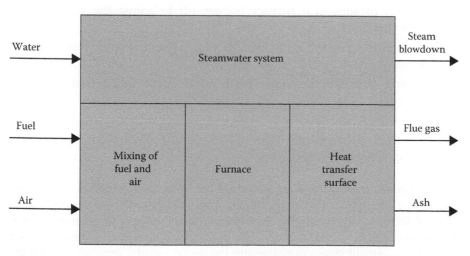

FIGURE 6.8
Basic block diagram of a boiler.

The fuel in the boiler is burnt to heat water and generate steam that drives the turbine. The turbine in turn drives the generator to produce electricity which is transmitted over transmission lines for consumer use. The production of electricity is unique in that electricity is consumed as soon as it is produced and can only be generated to the limit of consumer demand, which changes throughout the day. With the ever changing demand throughout the day, the power plant must generate enough power to match demand as soon as possible.

As the load varies, so does the system frequency. As the frequency changes, the speed governors of the turbines automatically adjust the turbine governor valve position to vary the steam flow to the turbines to support new load demands. The system frequency will stabilize once the balance between generation and consumption is produced. Since the generators meet the demand in a random manner, when the initial response is over, the system's load dispatcher brings the system frequency back to normal by sending out signals to the individual generating units. This signal is the load demand signal discussed in the boiler-control system. To summarize, the main control functions of the boiler are:

- Control of supply of fuel to boiler to adjust power and speed of steam turbine
- Control of fuel supply to the boiler to adjust the boiler drum pressure
- Regulation of feedwater pump speed to maintain the drum water level
- Control of induced draught fan speed to adjust the air pressure in the furnace
- Control of superheat furnace burner tilt to adjust the superheat steam pressure
- Control of superheat spray water flow to adjust the superheat steam temperature
- Control of reheat furnace burner tilt to adjust the reheat steam pressure
- Control of reheat spray water flow to adjust the reheat steam pressure
- Changing air damper position to control air flow through the boiler and exhaust gas temperature

The primary control loop regulates the fuel flow at the inlet to the boiler. Taking the example in Figure 6.9 [51], the fuel is regulated by a proportional-integral (PI) controller using the measured value of the output steam pressure. As the set point, the signal related to the desired output is given. The nonlinear block (o/p—output steam pressure and power set points) represents the relation between the power and the desired output steam pressure. During low loads, the boiler operates at a low pressure level in the constant pressure mode. When the load increases, the required pressure level increases in proportion

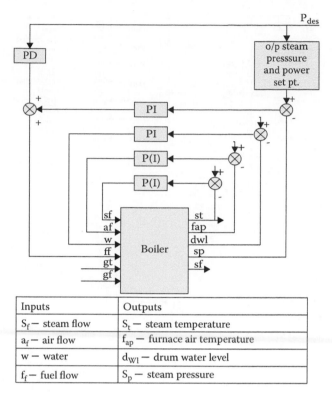

FIGURE 6.9
Boiler local control loops and related inputs and outputs.

to the desired output power until it attains a high-load constant value. The parallel feed forward proportional-derivative (PD) control path minimizes the control effort when the operators vary the required pressure.

The drum water level (d_{wl}) is an indication of the amount of water needed by the boiler. A PI controller, using the difference between the actual and desired drum water levels, regulates the water supply (w). The air flow to the boiler furnace (f_{ap}) is adjusted to maintain the air pressure (ap) in the furnace using a PI or P controller. The outlet steam pressure (s_p) is balanced by one or more water sprays to the superheat section of the boiler. During normal operating conditions, this loop plays a less significant role, providing only minor adjustments using a simple P or PI controller. The gas temperature (gt) or gas flow (gf) represents interconnection signals.

Figure 6.10 shows two proportional-integral-derivative (PID)-based control loops. The first loop is used to control furnace air pressure (ruG) by adjusting furnace input w_F (fuel flow to the furnace). The second loop is used to control the pressure of superheater steam (p_s) by adjusting air flow to the furnace (w_A). The third loop is used to control the drum water level (L) by adjusting feed water (w_e) from the economizer.

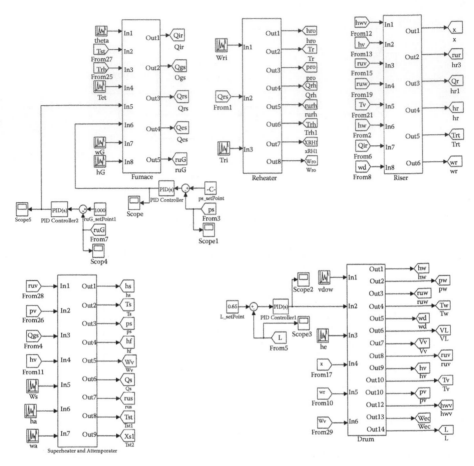

FIGURE 6.10
Model for power plant boiler-control system developed using MATLAB/Simulink.

6.4 Development of Steam-Temperature Control Using VisSim

The principles of energy conversation and mass balance are given in refs. [54, 55]. Normally, the steam temperature is controlled from both the fire and water side of a boiler. Feed forward combined with feedback-controlling strategies are widely applied in the steam-temperature control system. First, the fire side gas and water side steam dynamic models are developed in the project. Then all the models are integrated to simulate the process, including fuel combustion, gas flow, heat transfer, and steam dynamics. Finally, models using PI and PID controllers are developed to simulate the temperature control. VisSim 8.0 is used to simulate the dynamic steam-temperature control based on the models and the steam process and temperature control are simulated.

6.4.1 The Fire Side Process Simulation

The models include all processes from fuel and air blowing into the furnace to forming saturated steam. The combustion process principles and flue-gas dynamics models are given [51] in Equations 6.1 through 6.3. Figure 6.11 shows the gas temperature of the furnace model developed by VisSim 8.0.

$$h_{EG} = \frac{x_{F1}}{\rho_{EG}} \tag{6.1}$$

$$\frac{d}{dt}X_{F1} = \frac{1}{V_F}\left(C_F w_F + h_A w_A - Q_{ir} - Q_{is} - w_{EG}R_s\left(1+\frac{y}{100}\right)h_{EG}\right) \tag{6.2}$$

$$w_F + w_A + w_G - w_{EG} = V_F \frac{d}{dt}\rho_{EG} \tag{6.3}$$

where $X_{F1} = h_{EG} \times \rho_{EG}$ [J/m³], ρ_{EG} is the density of exhaust gas from the boiler [kg/m³], V_F the furnace volume [m³], w_F the fuel flow to the furnace [kg/s], h_A the inlet air enthalpy [J/kg], w_A the air flow to the furnace [kg/s], Q_{ir} the heat transferred to the risers [J/s], Q_{is} the heat transferred to the superheater [J/s], w_{EG} the mass flow of exhaust gas from the boiler [kg/s], R_s the stoichiometric air/fuel ratio, and y the percentage excess air [%].

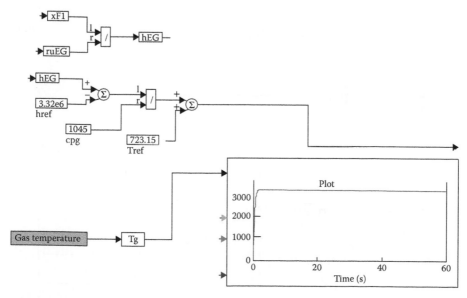

FIGURE 6.11
Model of the gas temperature of the furnace.

Figure 6.12 shows the model of the gas pressure of the furnace. Figure 6.13 shows the model of the heat transfer of the furnace, including heat transferred to the riser and the superheater by radiation, and heat transferred to the reheater and economizer by convection. The results of the heat transfer models in the furnace are shown in Figure 6.14.

6.4.2 The Water Side Process Simulation

The principles of heat and mass balance are given in refs. [51,54,57]. The drum water model, saturated steam model from drum to superheater, and steam model in the superheater are developed based on the principles (Equations 6.4 through 6.9). The water side process is simulated using VisSim 8.0 and the results are shown in Figures 6.15 through 6.18.

$$\frac{d}{dt}m_{DL} = w_e + (1-x)w_r - w_d - w_{ec} \tag{6.4}$$

$$\frac{d}{dt}X_{D1} = w_e h_e + (1-x)w_r h_{wy} - w_d h_w - w_{ec} h_y \tag{6.5}$$

$$\frac{d}{dt}X_{D2} = w_{ec} + x w_r - w_y \tag{6.6}$$

$$\frac{d}{dt}\rho_s = \frac{1}{V_s} + (w_y - w_s) \tag{6.7}$$

$$\frac{d}{dt}T_{st} = \frac{1}{M_s C_{st}} + (Q_{gs} - Q_s) \tag{6.8}$$

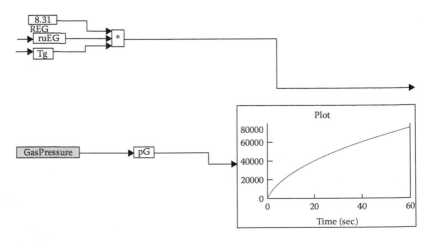

FIGURE 6.12
Model of the gas pressure of the furnace.

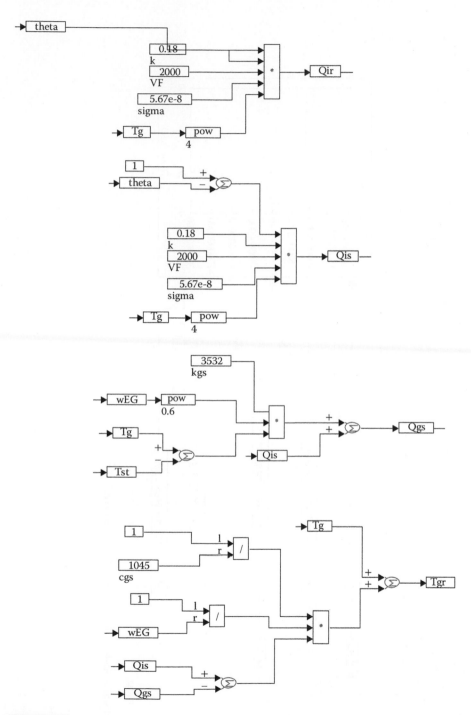

FIGURE 6.13
Models of heat transfer in a furnace.

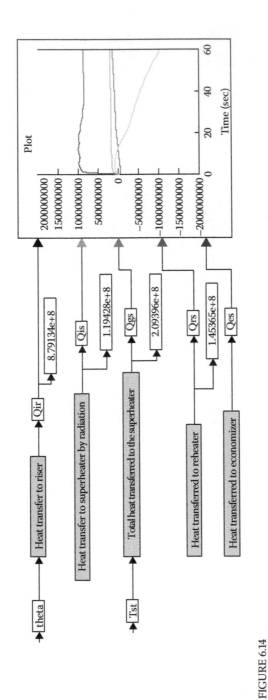

FIGURE 6.14

Results of models for heat transfer in the furnace.

$$\frac{d}{dt}X_{s1} = \frac{1}{V_s} + \left(Q_s + w_y h_y - w_s h_s + \left(h_a - h_f\right)w_a\right) \tag{6.9}$$

where m_{DL} is the drum liquid mass [kg], w_e is the water flow from the econo-mizer [J/kg], w_r is the liquid–vapor mixture mass flow from the risers [kg/s], w_d is the water mass down to the downcomer [kg/s], w_{ec} is the evaporation of water [kg/s], h_e is the specific enthalpy of water from the economizer [J/kg], h_{wy} is the enthalpy of steam flow to the superheater [J/kg], h_w is the enthalpy of drum water [J/kg], h_y is the drum outlet steam specific enthalpy [J/kg], w_y is the steam flow to the superheater [kg/s], V_s is the superheater volume [m³], M_s is the superheater mass [kg], C_{st} is the heat capacitance of super-heater tubes [J/(kg·K)], Q_{gs} is the heat flow from the furnace [J/s], Q_s is the heat transferred to the steam [J/s], w_s is the steam flowing from the superheater [kg/s], h_s is the specific enthalpy of superheated steam [J/kg], h_a is the specific enthalpy of attemperation water [J/kg], h_f is the specific enthalpy of evapora-tion [J/kg], and w_a is the water flow of attemporation [kg/s].

Figure 6.15 shows the results of the model of water in the drum.

Figure 6.16 shows the model of saturated steam in the superheater.

Figure 6.17 shows the results of the model of steam from the drum to the superheater.

Figure 6.18 shows the results of the model of saturated steam in the superheater.

6.4.3 Combining the Fire and Water Side Models

The steam properties of the fire side and water side are combined to create an integrated model as shown in Figure 6.19. Figure 6.20 shows the models that simulate steam-temperature control using a PID controller.

6.5 Simulation of Heat-Transfer Processes Using Comsol 4.3

6.5.1 Introduction

Improving the boiler-combustion process to increase the overall boiler efficiency is profitable to the power-generation industry. Slagging-related boiler-combustion problems still badly influence the power-generation industry by decreasing the heat-transfer efficiency of equipment inside the boiler and increasing carbon emissions. Although neural network–based methods have been applied to optimize the coal-fired power plant boiler-combustion process and increase the efficiency this approach is not always successful. For example, slagging and fouling accumulating on the surface of heat-transfer equipment or the heat-convection pass can not only deterio-rate the boiler-combustion efficiency but also lead to severe potential threats to the boiler [58]. It is difficult to apply a neural network–based method to restrict increases in slagging and fouling due to nonavailability of accurate measuring data regarding slagging and fouling status. With data from an instrument, the neural network can be trained to approximate highly non-linear functions, since the neural network depends on the input/output data but not on the physical structure of the system. The neural network–based method does not work successfully without instrumental data. The main reasons for slagging and fouling are boiler design and operation [59]. Moreover, boilers of identical design apparently firing identical fuels have often been reported to encounter quite different slagging and fouling problems [3], so there is a close relation between the behavior of the combustion process and boiler efficiency. Improving the fields of flue-gas properties such as the temperature field of the flue gas can increase boiler-combustion efficiency.

As CFD has been used to model the complex-combustion process, which has achieved successful assessment of boiler performance [13,60–64,65,66], this

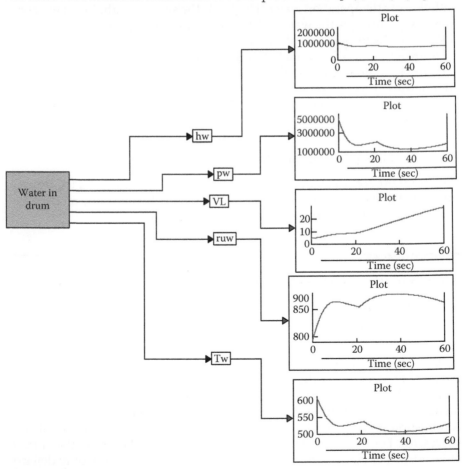

FIGURE 6.15
Results of the model of water in the drum.

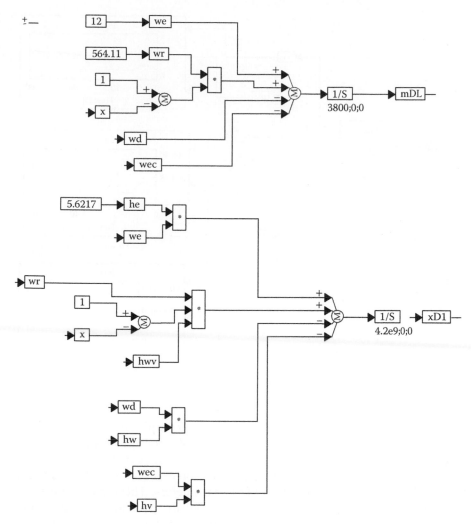

FIGURE 6.16
Model of saturated steam in the superheater.

research proposes a method of integrating a GA with CFD to improve the combustion process. The model can obtain the data of the boiler-combustion process such as flue-gas properties; these data are difficult to get normally.

6.5.2 A Simple Model of a Combustion Process with Heat-Transfer Efficiency Influenced by Slagging

CFD is a part of simulation technology that is used to forecast or reconstruct the behavior of an engineering product or physical situation under assumed or measured boundary conditions (geometry, initial states, load, etc.) [54]. It

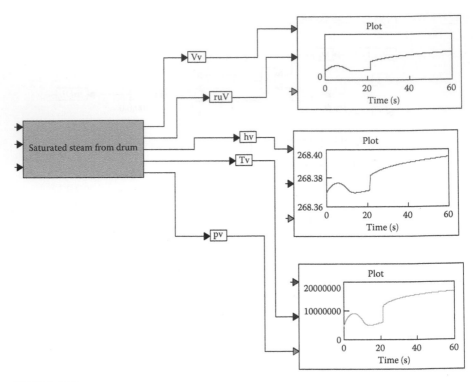

FIGURE 6.17
Results of model of steam from drum to superheater.

has been successfully applied to simulate highly complex industry processes [67–70]. This chapter proposes a simple model of a combustion process considering the influence of slagging on heat-transfer efficiency.

The dynamic system shown in Figure 6.21 consists of a fireball which is a heat source controlled by the two input parameters. These parameters are assumed to be fuel feeding speed and primary air speed at which the powder mix of fuel and air are blown into the furnace and become the heat source of the fireball. The fireball is inside a block where the top side is insulated and the other three sides are made of glass. A slagging layer is assumed and its conductivity is influenced by the input parameters. An arbitrary point is selected and the entire heat-transfer process is optimized and it is ensured that a high temperature exists at the selected point.

From the principles of CFD, a control volume (CV) is defined to be fixed in space and the fluid is assumed to flow through the CV, which is assumed to be located at (x_1, x_2, x_3) [71, 72].

The momentum equation of the system is given as

$$\frac{\partial(\rho_m u_i)}{\partial t} + \frac{\partial(\rho_m u_j u_i)}{\partial x_j} = \frac{\partial}{\partial x_j}\left[\mu_{\text{eff}}\frac{\partial u_i}{\partial x_j}\right] - \frac{\partial p}{\partial x_i} + \rho_m B_i + S_{u_i} \qquad (6.10)$$

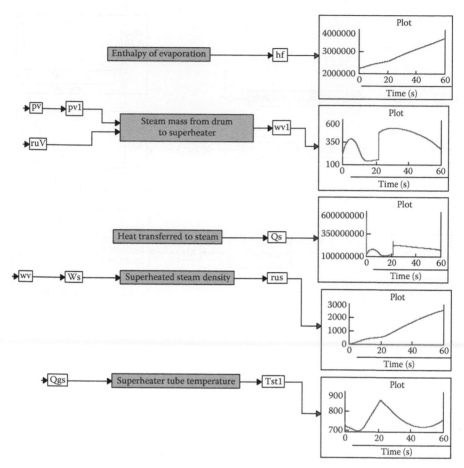

FIGURE 6.18
Results for model of saturated steam in superheater.

where ρ_m is the density of the fluid (kg/m³), x_i, x_j are the displacements (m) in the i and j directions, u_i, u_j are the velocities (m/s) in the i and j directions, μ_{eff} is the viscosity of the fluid (kg/m s), B_i is the body force (N/kg), and S_{u_i} is the other source of momentum (kg·m/s).

The equation of mass transfer can be written as

$$\frac{\partial(\rho_m w_k)}{\partial t} + \frac{\partial(\rho_m u_j w_k)}{\partial x_j} = \frac{\partial}{\partial x_j}\left[\rho_m D \frac{\partial w_k}{\partial x_j}\right] + R_k \tag{6.11}$$

where $w_k = \dfrac{\rho_k}{\rho_m}$ is the mass fraction, $\displaystyle\sum_{\text{all species}} w_k = 1$, D is the mass diffusivity (m²/s), and R_k is the rate of generation in the CV.

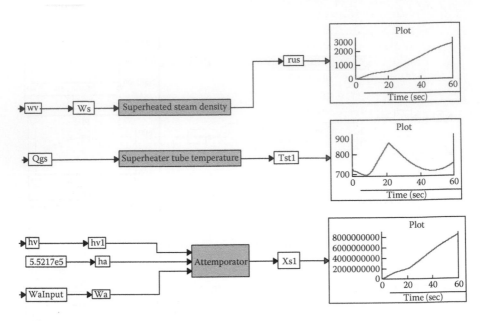

FIGURE 6.19
Model integration for the fire and water sides.

The energy equation of the system is given as

$$E = Q_{conv} + Q_{cond} + Q_{gen} - W_s - W_b \tag{6.12}$$

where E is the rate of change of energy of the CV (J/s),
Q_{conv} is the net rate of energy transferred by convection (J/s),
Q_{cond} is the net rate of energy transferred by conduction (J/s),
Q_{gen} is the net volumetric heat generation within the CV (J/s),
W_s is the net rate of work done by surface forces (J/s),
W_b is the net rate of work done by body forces (J/s).

In the specific case, assume the heat flux perpendicular to the surface of a CV is indicated by the terms Q_x'', Q_y'', and Q_z''. The heat flux at the opposite surfaces can then be expressed using the first-order Tayler series expression, as follows:

$$Q_{x+dx}'' = Q_x'' + \frac{\partial Q_x''}{\partial x} dx \tag{6.13}$$

$$Q_{y+dy}'' = Q_y'' + \frac{\partial Q_y''}{\partial y} dy \tag{6.14}$$

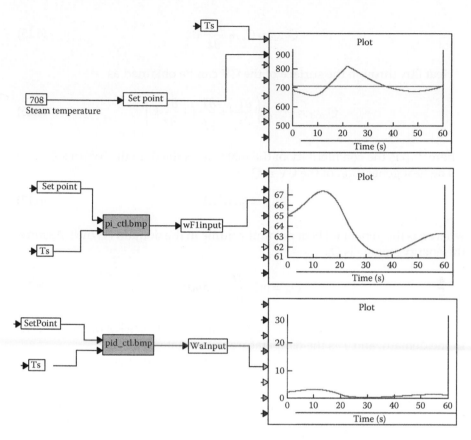

FIGURE 6.20
Steam-control models and simulation results.

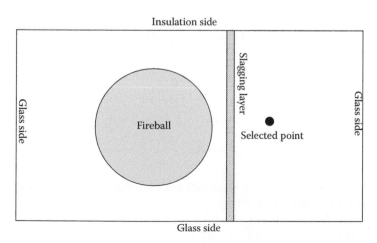

FIGURE 6.21
The geometry of the heat-transfer process with a slagging layer.

$$Q''_{z+dz} = Q''_z + \frac{\partial Q''_z}{\partial z} dz \tag{6.15}$$

Heat flux through the surface of the CV can be obtained as

$$Q_{cond} = -k_{cond} \left(\frac{\partial T}{\partial x} i + \frac{\partial T}{\partial y} j + \frac{\partial T}{\partial z} k \right) \tag{6.16}$$

where k_{cond} is the conductivity of the material in the domain (W/(m·K)). The heat generation of the CV is Q_{gen}:

$$Q_{gen} = q dx dy dz \tag{6.17}$$

where q is the generated heat per unit volume in the domain (W/m³). Assume the energy storage is Q_{st}:

$$Q_{st} = \rho C_p \frac{\partial T}{\partial t} dx dy dz \tag{6.18}$$

where ρ is the density (kg/m³), C_p is the specific heat (J/(g·K)) of the material in the domain, and T is the temperature (K) of material.

Substituting Equations 5.7 through 5.9 into 5.3, the energy conservation equation becomes

$$q dx dy dz + \left(\frac{\partial Q''_x}{\partial x} dy dz + \frac{\partial Q''_y}{\partial y} dx dz + \frac{\partial Q''_z}{\partial z} dx dy \right) = \rho C_p \frac{\partial T}{\partial t} dx dy dz \tag{6.19}$$

Q''_x, Q''_y and Q''_z can be obtained from Fourier's Law, as follows:

$$Q''_x = -k_{cond} \frac{dT}{dx} dx \tag{6.20}$$

$$Q''_y = -k_{cond} \frac{dT}{dy} dy \tag{6.21}$$

$$Q''_z = -k_{cond} \frac{dT}{dz} dz \tag{6.22}$$

Then the conduction energy equation per unit volume in Cartesian coordinates can be expressed as follows:

$$\frac{\partial}{\partial x} \left(k_{cond} \frac{\partial T}{\partial x} \right) + \frac{\partial}{\partial y} \left(k_{cond} \frac{\partial T}{\partial y} \right) + \frac{\partial}{\partial z} \left(k_{cond} \frac{\partial T}{\partial z} \right) + q = \rho C_p \frac{\partial T}{\partial t} \tag{6.23}$$

When the system reaches a steady-state condition, the term $\dfrac{\partial T}{\partial t}$ is elimi-nated. If the thermal conductivity is independent of the direction, the con-duction energy equation can be written in a simple form as follows:

$$\frac{\partial^2 T}{\partial x^2} + \frac{\partial^2 T}{\partial y^2} + \frac{\partial^2 T}{\partial z^2} + \frac{q}{k_{\text{cond}}} = \frac{\rho C_p}{k_{\text{cond}}} \frac{\partial T}{\partial t} \tag{6.24}$$

The boundary condition comes from the side around the system. The con-stant temperature T_s, also called the Dirichlet condition, corresponds to a situation for which the surface is maintained at a fixed temperature at all times. The boundary condition is as follows:

$$T(x,t) = T_s \tag{6.25}$$

The second boundary condition, also called the Neumann condition, corresponds to a constant heat flux applied to a surface. The heat flux q_s'' is related to the temperature gradient at the surface by Fourier's Law,

$$-k_{\text{cond}} \frac{\partial T}{\partial x} = q_s'' \tag{6.26}$$

A special case of the Neumann boundary condition is an insulated bound-ary condition, and the heat flux can be zero:

$$-k_{\text{cond}} \frac{\partial T}{\partial x} = 0 \tag{6.27}$$

The third boundary condition corresponds to convection at a surface. The conduction-convection heat balance at the wall surface must be satisfied and the heat-transfer coefficient h and the exterior temperature of the boiler T_∞ should be known:

$$-k_{\text{cond}} \frac{\partial T}{\partial x} = h\left[T_\infty - T(x,t)\right] \tag{6.28}$$

The finite element method (FEM) is utilized to solve governing Equations 6.23 or 6.24 with the boundary condition equations 6.25 through 6.28 and to discretize the computational domain. A linear triangle element is selected for two-dimensional analysis. Assume the nodes of an element are named i, j, and m. At each node, there are two degrees of freedom, displacement in x and y. The temperature at the nodes T_i, T_j, and T_m are expressed in the fol-lowing matrix form:

$$\{T\} = \begin{bmatrix} N_i, & N_j, & N_m \end{bmatrix} \begin{Bmatrix} T_i \\ T_j \\ T_m \end{Bmatrix} \tag{6.29}$$

where N_i, N_j, and N_m are linear shape functions given by

$$N_i = \frac{1}{2A}(\alpha_i + \beta_i x + \gamma_i y) \tag{6.30}$$

$$N_j = \frac{1}{2A}(\alpha_j + \beta_j x + \gamma_j y) \tag{6.31}$$

$$N_m = \frac{1}{2A}(\alpha_m + \beta_m x + \gamma_m y) \tag{6.32}$$

The expressions for α_s, β_s, and γ_s ($s = i$, j, and m) are defined as follows:

$$\alpha_i = x_j y_m - y_j x_m \qquad \alpha_j = x_m y_i - y_m x_i \qquad \alpha_m = x_i y_j - y_i x_j$$

$$\beta_i = y_j - y_m \quad \beta_j = y_m - y_i \quad \beta_m = y_i - y_j \tag{6.33}$$

$$\gamma_i = x_m - x_j \qquad \gamma_j = x_i - x_m \qquad \gamma_m = x_j - x_i$$

The temperature gradient matrix is given as follows:

$$\{g\} = \left\{ \begin{array}{c} \dfrac{\partial T}{\partial x} \\[2mm] \dfrac{\partial T}{\partial y} \end{array} \right\} \tag{6.34}$$

The heat flux and temperature gradient are related to each other using the thermal conductivity matrix $[D]$ as follows:

$$\left\{ \begin{array}{c} g_x \\ g_y \end{array} \right\} = -[D]\{g\} \tag{6.35}$$

and the thermal conductivity matrix $[D]$ is defined as

$$[D] = \left[\begin{array}{cc} k_{xx} & 0 \\ 0 & k_{yy} \end{array} \right] \tag{6.36}$$

Substituting Equation 6.29 in Equation 6.34, we have

$$\{\underline{g}\} = \begin{bmatrix} \dfrac{\partial N_i}{\partial x} & \dfrac{\partial N_j}{\partial x} & \dfrac{\partial N_m}{\partial x} \\[2mm] \dfrac{\partial N_i}{\partial y} & \dfrac{\partial N_j}{\partial y} & \dfrac{\partial N_m}{\partial y} \end{bmatrix} \begin{bmatrix} T_i \\ T_j \\ T_m \end{bmatrix} \qquad (6.37)$$

The temperature gradient matrix can also be written in a compact form as

$$\{\underline{g}\} = [\underline{B}]\{T\} \qquad (6.38)$$

The $\underline{B}$ matrix is defined as

$$[\underline{B}] = \frac{1}{2A}\begin{bmatrix} \beta_i & \beta_j & \beta_m \\ \gamma_i & \gamma_j & \gamma_m \end{bmatrix} \qquad (6.39)$$

The stiffness matrix is obtained from the potential energy theory as follows:

$$[K] = \iiint_v [\underline{B}]^T [D][\underline{B}]\mathrm{d}V + \iint h[N]^T [N]\mathrm{d}s \qquad (6.40)$$

where the first term contributes for the conduction, while the second term contributes for convection. The element equation is formulated in the form of $\{f\} = [K]\{T\}$ and the force matrix represents heat flow at the element's boundary, and it is defined as

$$\{f\} = \frac{QAL + q''PL + hT_\infty PL}{2}\begin{Bmatrix} 1 \\ 1 \end{Bmatrix} \qquad (6.41)$$

where P is the perimeter of the element, A is the area perpendicular to heat flow, Q is the heat generation in the element, q'' is the heat flux at the boundary of the element, and L is the element's side length.

Finally, the physical domain of the case in this paper can be divided into a number of subdomains which include the slagging layer subdomain, heat source subdomain, and other heat transfer subdomains. The heat transfer in each subdomain can be calculated using an FEM. The model and features of the heat-transfer process with slagging in this case is created using commercial software Comsol 3.2a, which is an FEM-based CFD tool widely applied in the scientific research field and design industry. Figure 6.22 shows the model geometry of a heat-transfer process considering the slagging layer. The results of the model show the temperature distribution and heat flux of the heat-transfer process in the case depicted in Figure 6.22.

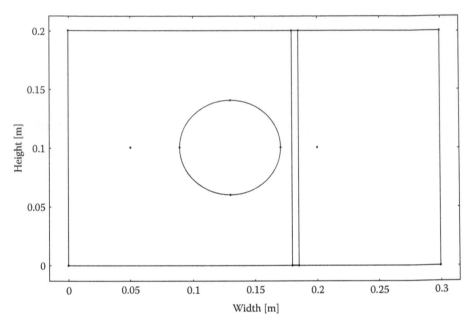

FIGURE 6.22
The model of a heat-transfer process with slagging.

First, Figure 6.23 shows that the temperature distribution is influenced by a slagging layer with a normal level of slagging deposition. The temperature distribution has a discontinuity between the left side and right side of the slagging layer. This is caused by decreasing heat conductivity of the slagging layer. Figure 6.24 shows a high level of slagging buildup. The slagging layer almost blocks the fireball heat transfer from the left side to the right side and there is a sudden massive temperature fall. Figure 6.25 shows a minor temperature fall when heat transfers from the left side to the right side of the slagging layer and less temperature difference exists between the two sides of the slagging layer.

It is observed from Figures 6.23 to 6.25 that the temperature of the selected point reaches the highest level when the slagging is low and it reaches the lowest level when the slagging is high. Normally, the temperature of a selected point falls within the range of the highest and lowest one as shown in Figure 6.24.

Second, the heat-flux distribution of the system has a similar trend as the temperature distribution. This can be found by comparing the two kinds of distribution in different conditions from Figures 6.26 to 6.28. Figure 6.26 shows that the heat-flux decreases when it transfers through the slagging layer. The slagging layer causes some loss of heat-transfer efficiency. Figure 6.27 shows that all heat flux is almost blocked by the slagging layer. The heat flux in the right side of the slagging layer nearly reaches zero. Figure 6.28 shows that the heat flux from the fireball transfers from the left side to the right side with little loss and the heat-transfer efficiency reaches

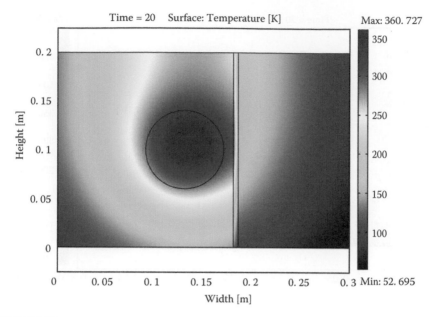

FIGURE 6.23
The temperature distribution with a normal level of slagging.

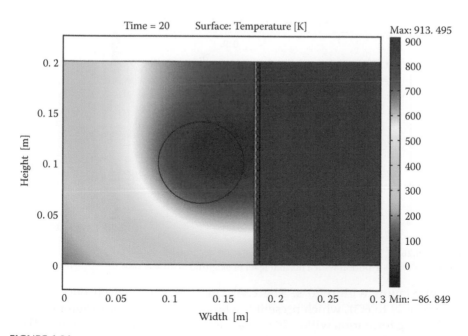

FIGURE 6.24
The temperature distribution with a high level of slagging.

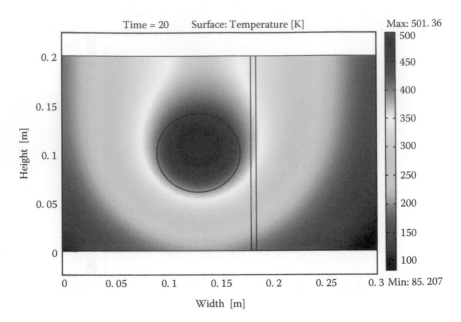

FIGURE 6.25
The temperature distribution with a low level of slagging.

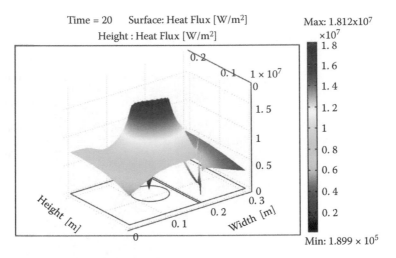

FIGURE 6.26
The heat-flux distribution with a normal level of slagging at 20 seconds.

the maximum within 20 s. The consequences can also be observed in Figures 6.29 to 6.31, which present the flux trending under different levels of slagging formation within 20 s. The maximum heat flux reaches 6.1 MW/m² in normal slagging conditions as shown in Figure 6.29, while it is only about 0.16 MW/m² with a higher slagging level as presented in Figure 6.30.

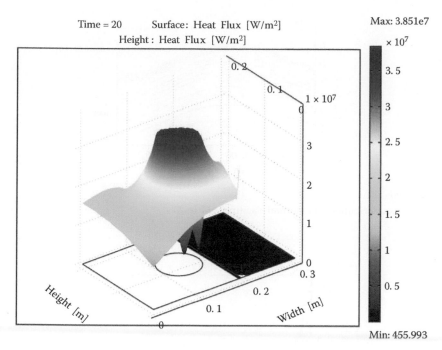

FIGURE 6.27
The heat-flux distribution with a high level of slagging at 20 seconds.

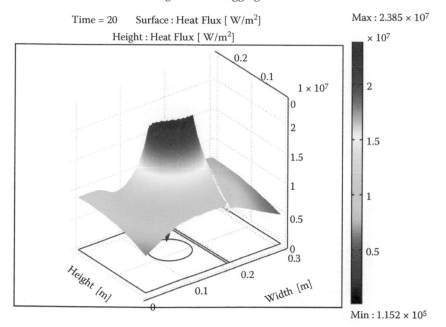

FIGURE 6.28
The heat-flux distribution with a low level of slagging at 20 seconds.

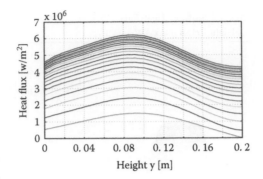

FIGURE 6.29
The heat-flux distribution with normal slagging within 20 s.

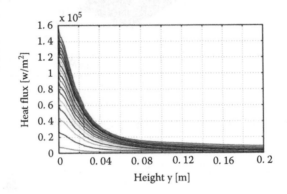

FIGURE 6.30
The heat-flux distribution with a higher slagging level within 20 s.

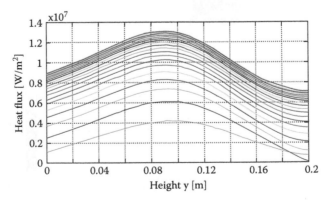

FIGURE 6.31
The heat-flux distribution with less slagging within 20 s.

TABLE 6.1

Comparison of Slagging Thickness and Loss of Heat Conductivity

Slagging Thickness (inch)	Loss of Heat Conductivity (%)
1/32	9.5
1/16	26.2
1/8	45.3
3/16	69

Figure 6.31 shows that the maximum heat flux of the system reaches 13 MW/m^2 with a lower slagging level.

Finally, the trending of the temperature of the selected point in Figures 6.23 through 6.28 shows that a higher temperature can be obtained at the selected point if input parameters are adjusted properly like in Figures 6.25, 6.28, and 6.31. This is the target of the research, which can be achieved using the proposed optimizing strategy presented in the chapter. In a real fossil fuel boiler-combustion process, with the combustion conditions changing, conventional combustion control is found to be insufficient to optimize the combustion process and cannot control slagging formation, which can effectively impair boiler efficiency. In power-generation industries boiler running, a same fireball is created by turbulence of coal powder, primary air, and secondary air. Similarly, a slagging layer can be formed on the surface of the water wall, superheaters, repeaters, and other heat-transfer equipment because of the variation of coal quality and inappropriate operation. Table 6.1 illustrates the loss of heat conductivity due to slagging accumulation on the surface of heat-transfer equipment [73].

In the heat-transfer model, the adherence of slagging to the surface of heat-transfer equipment can lead to a severe loss of heat-transfer efficiency. On the other hand, the optimal set point values for conventional-controller parameters should exist and can be found to optimize boiler combustion.

6.5.3 Creating a GA Model and Validating It Using Simulink

GA simulates evolution and is best viewed as a type of global-optimization process [74–79]. For Figure 6.21, GA is applied to optimize the heat-transfer process and the higher temperature is obtained at the selected point. A GA model is used to optimize a heat-transfer process. In addition, a real-time process model is created using Simulink to validate the proposed GA method. The thermal condition of the slagging layer is assumed to have a relation with the two input parameters, Heater1 and Heater2, as shown in Figure 6.32.

Based on the heat-transfer process in this case, the GA and its instance are given as follows [80]:

$$\theta^j(k) = [\ \theta_1^j(k),\ \theta_2^j(k),\ldots,\ \theta_p^j(k),]^T \tag{6.42}$$

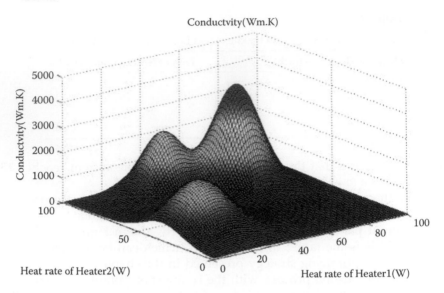

FIGURE 6.32
The relationship between the slagging layer conductivity and inputs.

where the generation number is k, $\theta_i^j(k)$ is a single parameter at time k, j refers to the jth chromosome, and $\theta_i^j(k)$ refers to the ith trait on the jth chromosome. Suppose that chromosome j is composed of P of these parameters (traits). The population of individuals at time k is given by

$$P(k) = \left\{ \theta^j(k) : j = 1, 2, \ldots, S \right\} \qquad (6.43)$$

where the number of individuals in the population is S. The population $P(k)$ at time k often refers to the generation of individuals at time k. Basically, according to Darwin, the most qualified individuals survive to mate and produce offspring. It is assumed that the individual fitness is $\bar{J}\left(\theta^j(k)\right)$ at time k. The online individual-fitness value is obtained from the strategy on combining the GA with CFD.

Genetic Operations: Chromosome, crossover, and gene mutation happen in a mating pool $M(k)$ where the most qualified individuals are selected:

$$M(k) = \left\{ m^j(k) : j = 1, 2, \ldots, S \right\} \qquad (6.44)$$

In this case, the roulette wheel selection strategy is applied to select an individual for mating. An individual $m^j(k)$ is selected equally to $\theta^j(k) \in P(k)$ with probability

$$p_i = \frac{\bar{J}\left(\theta^j(k)\right)}{\sum_{J=1}^{S} \bar{J}\left(\theta^j(k)\right)} \qquad (6.45)$$

Figure 5.12 shows the relationship between traits (inputs of the system) and fitness (the assumed heat flux of slagging layer).

Figure 6.33 shows a model developed in Simulink to validate the instance of GA, in which the module GA-optimization method has two inputs that are Heater1 and Heater2 as shown in Figure 6.32. It also has several outputs; one output is GAOutput_u1, which stands for the heat conductivity as shown in Figure 6.32. GA is the control strategy of the model. At the same time, the function of the assumed relationship between the inputs Heater1 and Heater2 which lead to slagging formation and heat conductivity is included in the model. Once the model is running, it can find out the optimal parameter values for Heater1 and Heater2, which maintain much less slagging accumulation and a higher conductivity for heat-transfer equipment inside the boiler. After 100 generations of the GA, a pair of input parameter values for Heater1 and Heater2 are found. The process of using GA to find out optimal parameter values is shown by contour in Figure 6.34, where the small circle with a star corresponds to the last pair of optimal values located in the domain for Heater1 and Heater2 after 100 generations of the GA. Other small circles shown in the contour plot correspond to values generated in the iteration process of the GA. In addition, it shows that the optimizing strategy successfully obtains a pair of input parameters that can create higher conductivity. Comparing Figures 6.32 and 6.34, the results are found to be very consistent.

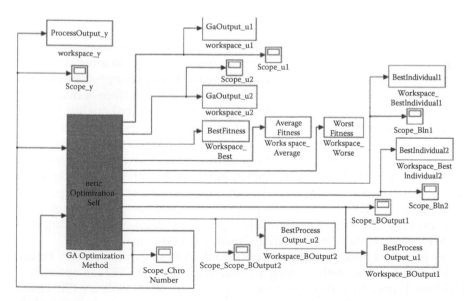

FIGURE 6.33
Validating the GA in a real-time process.

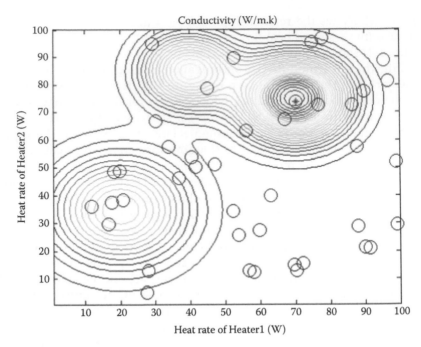

FIGURE 6.34
Input parameter trajectory on the contour plot of the temperature of a selected point.

6.5.4 Integrating a GA with CFD to Optimize the Heat-Transfer Process in Boiler Combustion

PID controllers are widely applied in control of the boiler-combustion process [81]. However, this conventional-control strategy is not able to achieve the best control performance when conditions inside the boiler change. For example, slagging may initiate and continuously deposit on the surface of heat transfer in the boiler because of variable parameters such as coal quality, blow of primary or second air, fineness of coal or coal feeder speed change. These changes in the combustion process may not be identified by PID controllers and may cause more heat loss and carbon emissions. The work in this research proposes a new method based on integrating a GA with CFD to adaptively tune the conventional PID controllers and optimize the combustion process. By combining a GA with CFD, the proposed method can prevent more slagging buildup and keep the combustion process continuously running in a more efficient mode when conditions inside the boiler are changing.

The proposed boiler-combustion-optimization strategy presented in Figure 6.35 shows how the method using an integrated GA and CFD can optimize a PID-based conventional-control process. The controller module is a PID-based conventional controller which sends controlling information u to the module named Combustion Process, which is a real-time boiler-combustion process

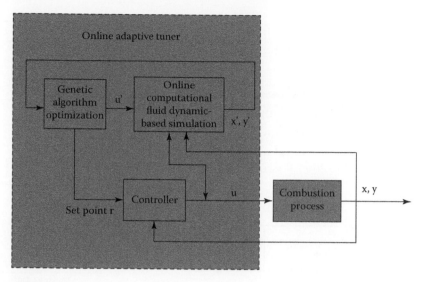

FIGURE 6.35
The logic flow of integrating a GA with CFD to optimize the combustion process.

with output x and y. The online CFD-based simulation obtains the multi-input parameters u, x, and y from the real-time process. This module not only simulates the general macroscopical real-time combustion process but also simulates microcosmic change in the combustion process, such as fields of flue-gas properties which are not controlled in a PID-based conventional-control system. However, these microcosmic characters of the process can cause a severe loss of boiler-combustion efficiency. The module output variables x' and y' are sent to the GA optimizing module, which applies the GA to find the optimal input parameter value u' for the online simulation module. The optimal value of u' is sent to the conventional controller as set point values. The conventional controller can achieve the most effective performance and improve the heat-transfer efficiency with optimal input set point values.

Based on the simple case of heat transfer in the combustion process with slagging buildup considered in Figure 6.21, the proposed optimization strategy is simulated using the Simulink model shown in Figure 6.36.

Figure 6.36 shows how to apply a GA to optimize a real-time industry process where a mathematical object function is unknown. In this specific case, the individual fitness function $\overline{J}\left(\theta^{j}(k)\right)$ is unknown. However, comparing with Figure 6.33, the individual fitness function $\overline{J}\left(\theta^{j}(k)\right)$ is obtained by an online simulation module rather than a function. The fitness values for each individual in a group of GAs have been obtained by online simulation of a heat-transfer process based on CFD.

First, the GA-optimization module sequentially sends 20 pairs of parameters to the combustion process module input parameters Heater1 and Heater2. Shortly, 20 outputs are fed back to the GA-optimization module in

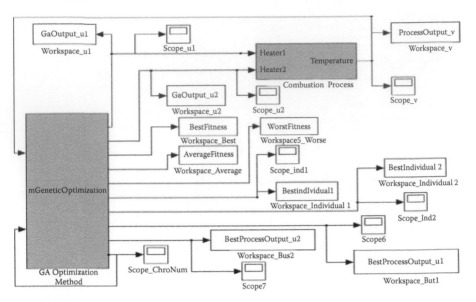

FIGURE 6.36
Integrating a GA with CFD is simulated using Simulink.

the same sequence. Using this, the GA-optimization module of obtains the fitness values for all individuals. Then, it can process the data in the current generation of the GA.

Second, a new sequence of input parameters is prepared and sent to the module of the combustion process. The iteration will not stop until 100 generations of GA are produced in this case.

Finally, the optimal input parameters are obtained and sent to the real-combustion process, which is assumed to be the same as the online-simulation module of the combustion process. Figure 6.37 presents the trend of input parameters within 100 generations of the GA. The result is also presented by lines produced by the best pair of input parameters at each iteration within 100 generations of the GA as shown in Figure 6.38, which shows that the proposed optimization strategy can find the optimal input parameters early in the 100 generations at this instance of the GA. The optimal parameters are input to the real combustion process, which is simulated using the Simulink model as shown in Figure 6.39.

The controller, using reasonable random input and without the proposed optimizations, is also simulated using the Simulink model and the results are presented in Figure 6.40. It is shown that the maximum temperature of the selected point is less than 300 K during the 700 s long test period.

Figure 6.40 also shows that the selected point obtained an encouraging higher temperature with the proposed optimizing strategy than that one without optimizing. The maximum temperature of the selected point is less than 300 K during the 700 s long test period if the model process is not

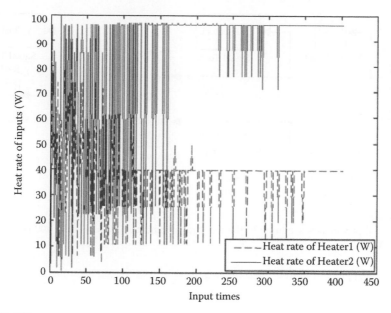

FIGURE 6.37
The input parameters within 100 generations of the GA.

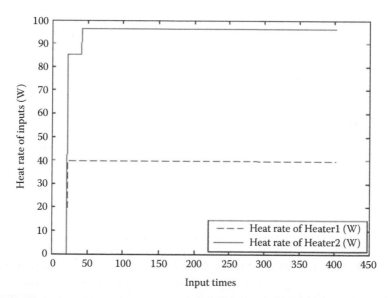

FIGURE 6.38
The optimal input parameters found within 100 generations of the GA.

optimized by the proposed optimizing strategy. However, after the first-time optimization with 100 generations of the GA made by the proposed optimization strategy, the maximum temperature of the selected point reaches

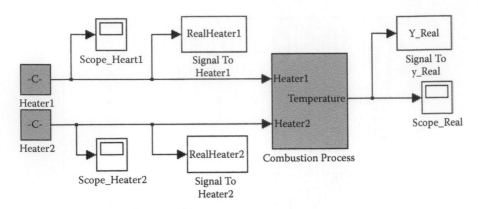

FIGURE 6.39
The optimal parameters are input to the real-time combustion process after 100 generations of the GA.

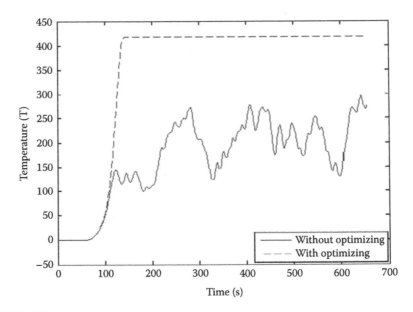

FIGURE 6.40
The comparison of temperature without and with optimizing after 100 generations of the GA.

about 420 K, which is much higher than the maximum temperature in the condition without the proposed optimization. In fact, the proposed online-optimization method can continuously tune real PID controllers by providing the optimal set point values with changing conditions inside the boiler. This can be presented by connecting the process running in Figure 6.36 to the process running in Figure 6.39 using the optimal input parameters.

This work has realized a new method by integrating a GA with a CFD model using Comsol to accurately simulate a real-time process and capture the optimal set point values to adaptively tune a PID controller. By tuning, the widely applied conventional controller can effectively restrict the slagging buildup in the combustion process using optimal-control parameters provided by the proposed method. Both the combustion heat transfer and adaptive tuning processes are simulated in this chapter and the results show that the optimum set point values can be found for a conventional controller to decrease slagging, which causes heat loss and thus increases heat-transfer efficiency.

6.6 Modeling the Combustion Processes in a Coal-Fired Power Plant Boiler Using ANSYS 14.5 and ANSYS Fluent 14.5

Coal-fired power plant boiler combustion is a highly complex process, and improving the combustion process requires a method of multiobjective optimization. A combustion process with two objectives is shown in Figure 5.1, where Q_1 is the total heat absorbed by the tubes of heat-transfer equipment inside the boiler such as the water wall, superheater, reheater, and economizer, and maximum Q_1 is expected. Normally, if the temperature of the flue gas is higher than $T_{melting}$, which is the ash-melting temperature and is shown in Figure 5.1, the ash starts to melt and slagging increases. Therefore, an efficient boiler-combustion process should have maximum Q_1 with flue-gas temperature in the areas close to the sides of the furnace under $T_{melting}$.

Artificial intelligence (AI) technologies such as neural network–based methods and multiobjective optimization have been applied in industry to improve the efficiency of control systems [8,9,24,27,82–85]. For example, neural-network-driven computer systems are used to optimize soot-blowing in a coal plant boiler, reduce NO_x emissions, improve heat rate and unit efficiency, and reduce particulate matter emissions in coal-fired power plants in the United States [5]. Non-dominated sorting genetic algorithm II (NSGA II) is one of the AI-based multiobjective optimizations and it is widely used to successfully optimize industry processes [86,87–93]. In addition, CFD simulation technology is widely applied in the power-generation industry to analyze combustion processes [94,95], improve boiler design [96–100], and adjust burner tilt angle in an offline fashion after an overhaul or upgrade in a power plant [101].

In fact, with the advancement of computer technology and mathematical methodology, integrating AI with CFD technologies can solve combustion-related problems. Based on this, the research proposes new methods to improve combustion-process efficiency and decrease carbon emissions for the fossil fuel power-generation industry by integrating multiobjective optimization with CFD technology to improve boiler-combustion efficiency and decrease or even prevent serious slagging inside the furnace of a coal-fired power plant boiler.

The description of a CFD-based model of a coal-fired power plant boiler-combustion process is shown in Figure 5.1. Equation 5.18 gives an expression of the heat balance in the coal boiler-combustion process of Figure 5.1. Q_1 is normally in the range of 75%–90% [102]. However, slag accumulating on the heat-transfer surface can seriously influence Q_1. The thermal efficiency of the coal-fired boiler can be expressed as [102]

$$\eta = \frac{Q_{steam}}{Q_{coal}} \qquad (6.46)$$

where Q_{steam} is the useful heat out in steam and Q_{coal} is the heat in from coal:

$$Q_{coal} = Q_1 + Q_2 + Q_3 + Q_4 + Q_5 + Q_6 \qquad (6.47)$$

where $Q_1 - Q_6$ are the same as denoted in Equation 5.18.

The fields of temperature, pressure, velocity, and density of flue gas inside a coal boiler are dynamic and Equations 6.1 through 6.9 cannot be used to predict all the fields of flue gas accurately. Therefore, FEM supported with CFD is applied to simulate all the dynamic fields of the flue gas more accurately [21,20,81,54,22].

A three-dimensional power plant boiler furnace model is developed using ANSYS Fluent 14.5 based on real data, which is from a 1160 t/h tangential coal-fired power plant [54]. The characteristics of flue-gas property fields such as temperature and intensity are analyzed in the following and the results show that the simulation results of the flue-gas properties are close to the corresponding data from the power-generation industry and simulation results from research [21, 81]. The geometry model is developed based on the data from the power plant [54]. This is a 14.62 m wide, 12.43 m deep, and 48.8 m high furnace of the tangential-combustion type. The geometry of furnace is shown in Figure 6.41.

The positions of each set of four burners located in the same horizontal section are shown in Figure 6.42. The center line of the burner which is installed in a different corner is shown in Figure 6.43. The mesh for the geometry is shown in Figure 6.44. CORBA C++ is used to integrate ANSYS Fluent 14.5 with a multiobjective-optimization model developed using MATLAB. The details of the CFD-based coal-fired boiler-combustion model such as geometry data and combustion model equations and data can be studies in Section 4.3 of Chapter 4.

6.7 How to Integrate the Boiler, Turbine, and Generator System

The energy conservation and mass balance–based principles are given in refs. [54,55]. Based on these physical principles, the high-pressure section of the turbine model and generator model have been created using Equations 6.44

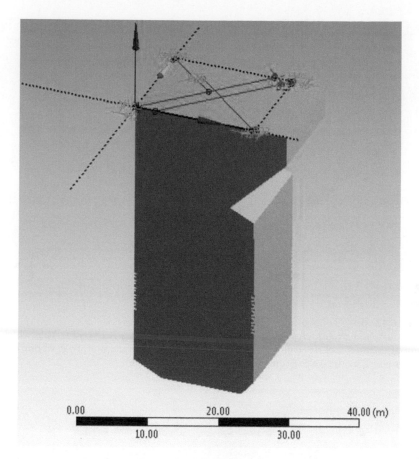

FIGURE 6.41

The geometry of a boiler of a coal-fired power plant developed using ANSYS DesignModeler 14.5.

through 6.47. Combining the boiler-control system models and steam-temperature-control model, an integrated model of boiler-turbine-generator has been developed which has a number of control loops, for example, steam-temperature control and steam-pressure control, and so on [103–106]:

$$w_i - w_{ohp} = V \frac{d}{dt} \rho_o \tag{6.48}$$

$$\frac{d}{dt} w_{ohp} = \frac{1}{\tau_{hp}} \left(w_i - w_{ohp} \right) \tag{6.49}$$

$$w_i h_i - w_{ohp} h_o = V \rho_o h_o \tag{6.50}$$

$$\frac{d\omega}{dt} = \frac{\omega_0}{2H_{eq}} \left[P_{mech} - P_{el} - D_{eq} \left(\omega - \omega_0 \right) \right] \tag{6.51}$$

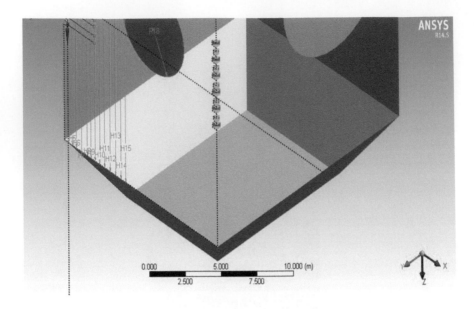

FIGURE 6.42
The burners and probes created in the geometry model using ICEM CFD 14.5, which is advanced analysis tool for geometry acquisition, mesh generation, and mesh optimization, and is provided by the ANSYS company.

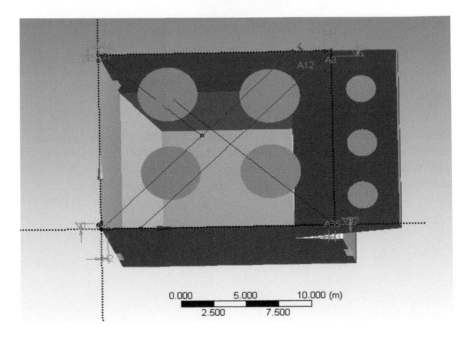

FIGURE 6.43
The center line of each burner viewed from the top of the furnace.

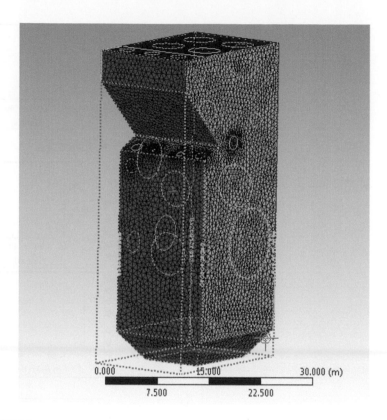

FIGURE 6.44
The mesh of the geometry model using ICEM CFD 14.5.

where w_i is the inlet steam flow from the boiler superheat section [kg/s], w_{ohp} is the outlet steam flow [kg/s], V is the high pressure (HP) section steam storage volume [m³], ρ_o is the steam density in the HP section steam storage [kg/m³], τ_{hp} is the HP section mass flow time constant [s], h_o is the outlet steam enthalpy [J/kg], ω is the rotor speed, ω_0 is the per unit reference rotor speed, H_{eq} is the initial constant for the machine, P_{mech} is the per unit mechanical shaft power, P_{el} is the per unit electrical load, and D_{eq} is the equivalent electrical damping coefficient.

6.8 Developing Models to Integrate the Boiler, Turbine, and Generator

6.8.1 Saturated Steam in the High-Pressure Section of a Turbine

Figure 6.45 shows the models and simulation results for the steam of the high-pressure section of a turbine.

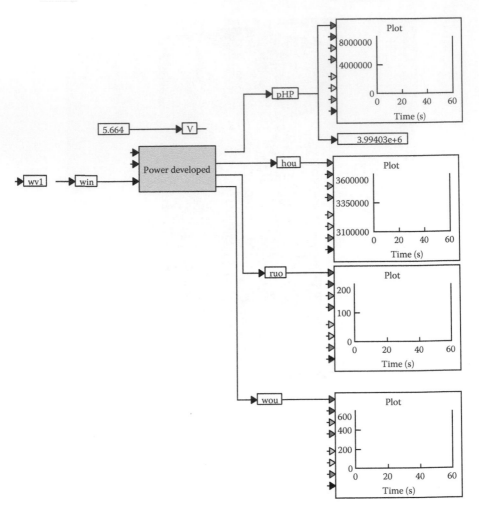

FIGURE 6.45
Model and simulation results for steam of the high-pressure section of a turbine.

6.8.2 Generator Models

Figure 6.46 shows the models and simulation results for the generator.

6.8.3 Integration of All the Models

All models are integrated based on the fuel or gas process and steam process by interlinking their inputs and corresponding outputs.

6.8.3.1 Connection of Furnace Fuel and Gas Model with Drum Model

The gas fuel and combustion process inside a furnace are combined to create the model shown in Figure 6.47.

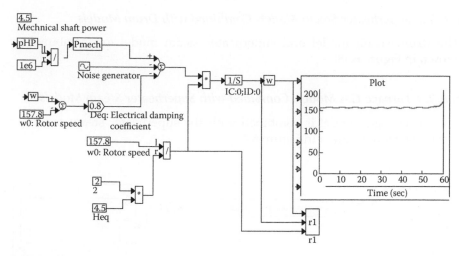

FIGURE 6.46
Model and simulation results for generator.

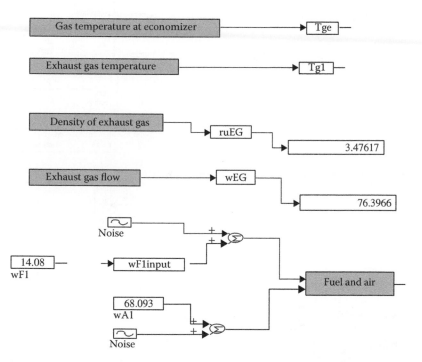

FIGURE 6.47
Combination of furnace fuel and gas process models with drum model.

6.8.3.2 Superheater Steam Models Combined with Drum Models

The drum steam model and superheater steam model are combined as shown in Figure 6.48.

6.8.3.3 Furnace Gas Models Combined with Superheater Steam Models

The furnace gas model is combined with the superheater steam model to create the model shown in Figure 6.49.

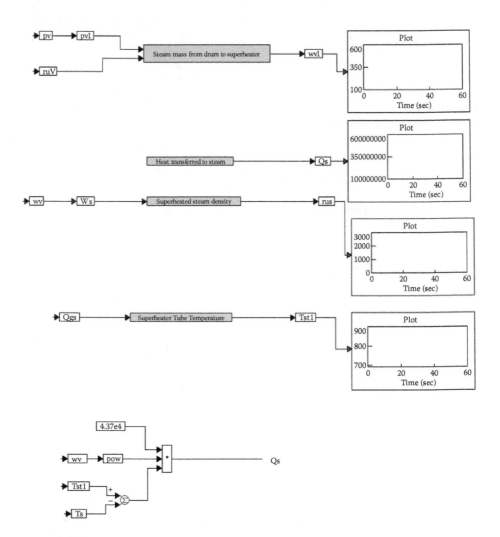

FIGURE 6.48
Combined drum steam model with superheater saturated steam through variables pv and ruV.

6.8.3.4 Superheater Steam Model Combined with High-Pressure Steam Model of Turbine

Superheater saturated steam models are combined with the steam process in the high-pressure section of a turbine as shown in Figure 6.50.

6.8.3.5 Control Model Integrated with Gas or Steam Process Models

PID controllers are used in the model to simulate keeping the fuel and air input and temperature at the proper values based on the power requirements from the generator. The model integration and encouraging results are shown in Figure 6.51.

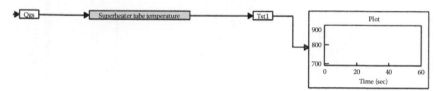

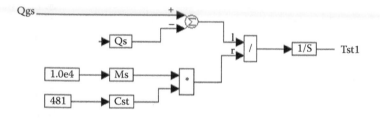

FIGURE 6.49
Furnace gas model combined with superheater steam model through variable Qgs.

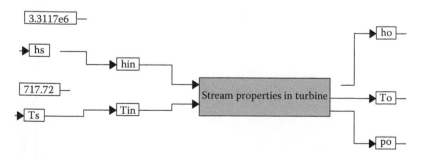

FIGURE 6.50
Combination of saturated steam with steam process in high-pressure section of turbine superheater model.

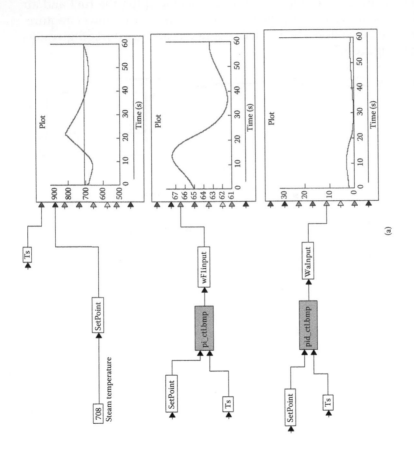

FIGURE 6.51

Adjusting the steam temperature based on Wa and wF1 in boiler system shown in (a). Adjusting the turbine and generator systems based on variable pHP shown in (b). Boiler, turbine, and generator systems are integrated through variables Ts and pHP.

(Continued)

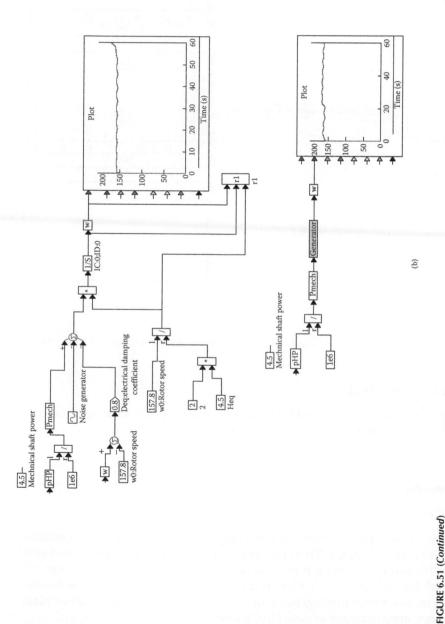

(b)

FIGURE 6.51 (Continued)
Adjusting the steam temperature based on Wa and wF1 in boiler system shown in (a). Adjusting the turbine and generator systems based on variable pHP shown in (b). Boiler, turbine, and generator systems are integrated through variables Ts and pHP.

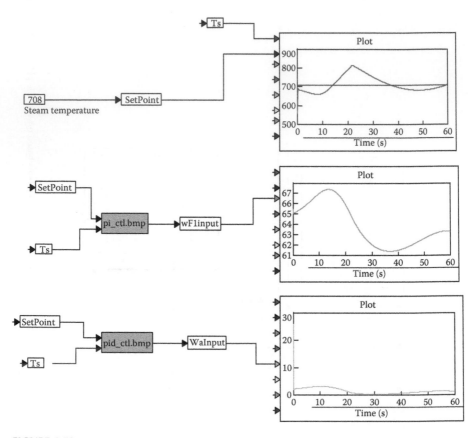

FIGURE 6.52
The control loop of adjusting air input and steam temperature to meet the power requirements from a generator.

6.8.3.6 *Control Models*

Figure 6.52 shows how PI and PID controllers are applied to control fuel and air to keep proper mechanical power to meet a load requirement from a generator, and encouraging results are obtained.

6.9 Conclusion

Thermal power plant processes, especially fossil fuel–fired boiler combustion, are highly complex. This chapter has discussed how to model and simulate thermal power plant processes using MATLAB/Simulink, VisSim 8.0, Comsol 4.3, and Ansys 14.5. More research methods concerning how to integrate computational intelligence with CFD to improve thermal power plant efficiency are discussed in detail in Chapters 7 through 9.

Part III

Thermal Power Plant Efficiency Improvement Modeling

7

Conventional Neural Network–Based Technologies for Improving Fossil Fuel Power Plant Efficiency

7.1 Introduction

Fossil fuel–fired boiler combustion optimization is significant for many countries because fossil fuel power plants still play a dominant role in electricity generation, and more than 40% of electricity worldwide is from fossil fuel. This research literature review, which focuses on how to solve the specific problems of the combustion process, shows that a number of methods are applied to improve fossil fuel boiler combustion efficiency. These methods improve the performance of conventional proportional–integral–derivative (PID)-based controllers, which are widely applied in the power-generation industry to optimize the combustion process. In addition, some of these methods are applied to assess the status of equipment of a boiler to tune the conventional control system by assigning an optimal set point value for parameters or to conduct operations to adjust the combustion system exactly and efficiently. Because the boiler-combustion process is a highly complex system with nonlinear, multi-input, and multioutput characteristics, these methods are based on neural networks (NNs), genetic algorithms, and digital-simulation technology.

7.2 NN-Based Power Plant Optimization Technology

Multilayer perceptron (MLP) NN is the most widely applied architecture for practical applications of NN, and normally consists of two layers of adaptive weights with full connectivity between inputs and hidden units, and between hidden units and outputs. The two-layer architecture is capable of universal approximation, which can approximate any continuous function to arbitrary accuracy.

An offline NN-based multivariable nonlinear controller is created to solve the problem caused by frequent load change in the thermal power plant with a highly competitive electricity market [24]. A multivariable long-range predictive control (LRPC) algorithm is applied to create a controller taking into account the problems from which conventional PID controllers suffer. An NN model with seven inputs, one hidden layer, and one output dynamic nonlinear MLP using a sigmoid activation function are formed. The NN model is trained through the multivariable controller and implemented on a detailed nonlinear boiler-turbine simulation. Then, two different cases of load changes are considered to test the NN-based controller in regulating the main steam temperature and main steam pressure. The results show that the NN-based controller gives superior performance in maintaining the main steam temperature, pressure, and reheat steam temperature within the recommended deviation ranges of load demand.

7.3 Online-Learning Applications

A MLP-based online-learning application is widely applied in the boiler control process. An online NN-based advanced controller is applied to improve the performance of boiler combustion [108]. The MLP is employed to build a predictive controller that obtains data from an optimization model which compares the difference between the set point value and boiler state estimator output as learning signals. Then, the gradient descent method is applied to optimize the network controller's weights, from which the nonlinear predictive control law under the reduced excess air level is derived. The NN-based advanced boiler-combustion controller is validated using a series of real-time data acquired from a running boiler system. Then it is tested by an offline simulation of the combustion process. The results are compared with the conventional PID controller on simulation and show that the NN-based advanced combustion controller has demonstrated the potency to improve boiler performance.

A radial basic function (RBF)-based NN is another kind of NN method which is suitable for online control [52]. It has shown the ability to approximate any continuous function at any precise value. In an RBF network, the mapping from the input layer to the hidden layer is nonlinear, but mapping from the hidden layer to the output layer is linear. Therefore, the learning speed of this kind of NN is significantly improved.

An NN model reference adaptive PID control method based on RBF NN is used to control the reheater temperature in a coal-fired power plant [52]. The NN model reference adaptive PID contains two sub-NNs. One is the neural network identifier (NNI), which is based on an RBF NN and is used to identify properties of online systems. The other is the neural network PID controller (NNC). With learning the data from the NNI, which identifies the controlled combustion process online, the NN model reference adaptive PID controller can adaptively control the online process by adjusting the weights of the NNC in a real-time fashion. The neural PID controller is tested by computer simulation and the results show that the NN can produce the necessary control signals to accomplish the control task and the performance of the neural PID controller is better than a conventional PID controller both in terms of stability and synthesizing performance.

NN indirect adaptive control with a fast-learning algorithm is applied to a real-time industry process in which using conventional instruments is difficult to obtain an exact mathematical model of the plant process because of the existing variations of parameters due to environmental disturbances and equipment aging. The method contains both NNI and NNC. Compared to the PID controller proposed in reference [52], the NNI is trained by the controller error instead of the estimated control error. In addition, a linear error signal is introduced to speed up the learning of the NNI and NNC. Therefore, the tracking capability and learning speed are outstanding and the controller is also robust against variations of the plant parameters.

An offline and online combined NN-based method is applied to control the process of an ultrasupercritical boiler [73]. First, offline NN-based controllers are created by training respectively from the four boiler control processes and are then combined to build a boiler NN combined model. Then a reference governor is applied to provide feed forward control action as well as the set point values to a feedback controller, and the feedback controller provides the actual control actions to the plant. A search algorithm is applied to find numerous candidate control actions and set point values that equally satisfy the cost function. With the cost function, a single set of control actions and set point values provide an optimal solution, or a reference governor can know which set is chosen. Intelligent gain tuning is done using an online identifier and a heuristic search to determine the gains of a PID control system. The heuristic search examines different gain values, and then simulates the system with these gain values and the online identifier. It continues to experiment with different gain values until it finds the set of gains that reduce the error between the set point value and the plant outputs. The results show that the controller is very effective for meeting the desired performance goals of the boiler system.

NN-based commercial software has been applied in the power-generation industry to improve the efficiency of boiler combustion. Using NN, model predictive control, and direct search technologies, NeuCo's CombustionOpt can determine the optimal fuel type and air set points for the specific goals and constraints and then make the necessary adjustments to the fuel and air variables in real time [4]. NN-based optimization technologies have been applied in United States power plant optimization demonstration projects to improve the fossil fuel power plant combustion process [5]. The NN algorithm learns the relationships between operating conditions, emissions, and performance parameters by a training process and develops a highly complex nonlinear function which maps the system inputs to the corresponding outputs. The data from this function are passed to the mathematical-optimization algorithm which can find optimum operating conditions.

Although NN-based technologies are employed to successfully solve problems in the power-generation industry, some problems still seriously impair the efficiency of heat transfer, degrading the performance of the boiler in a power plant. For example, slagging and fouling are still serious problems in power plants that use fossil fuels with high slagging and fouling tendencies. The slagging deposited on the surface of a water wall and fouling accumulated in the superheater not only badly reduce the heat-transfer efficiency but also cause unplanned outages and maintenance with huge expenditures. In this situation, the level of slagging and fouling is very difficult to measure if sufficient output data are not available to train the NN-based model to approximate the relationship between the input and output variables. In addition, without an approximation function, it is very difficult for conventional PID controllers to control the combustion process using the widely applied principles of PID strategy.

7.4 Finite Element Method–Supported Computational Fluid Dynamics (CFD) Technology Applications in Power Plant Boiler Simulation

Literature review shows a number of applications related to CFD in which the microscopic physical and chemical properties of the equipment are simulated using the equations of thermal dynamics, fluid dynamics, and chemical reactions. They have been implemented to assess the performance of the heat-transfer surface of equipment and simulate some heat-transfer processes such as slagging formation, erosion, and corrosion occurring on the surface of heat-transfer equipment.

A CFD-based slagging-deposition model for a 575 MW tangential-fired pulverized coal boiler-combustion process has been developed [109]. Ash particles are assumed to accumulate on the surface of water wall tubes only if they are in the boundary layer of the water wall. A particle with temperature T_p and velocity v with three components v_x, v_y and v_z is defined and only a particle with a nonzero perpendicular component of the velocity directed toward the water wall can adhere. Furthermore, with defining initial deformation temperature (IDT) and the flow temperature (FT), if IDT $< T_p <$ FT, then the particle is neither solid nor liquid and the viscous particle can adhere on the water wall surface with probability between zero to unity. If $T_p =$ IDT, then the probability for an ash particle to adhere on the water wall surface is zero. Whereas if $T_p =$ FT, then the probability for an ash particle to adhere is unity, where the temperature has the same dependence as the viscosity. Therefore, the ash particles are assumed to deposit on the surface of water wall tubes only if its temperature is in the flow-liquid range. CFD-based simulation determines the particle number density at the water wall boundary layer, particle temperature, and particle velocity vector. The model is assessed on the 575 MW tangential-fired boiler using two kinds of local coal and the results yield a good description of slagging and fouling in various sections of the boiler.

Another CFD-based combustion process simulation is integrated with the ash behavior tool, AshPro^SM, which is developed by the University of North Dakota Energy and Environment Research Center, USA, to predict slagging formation on the heat-transfer surface [110]. The tube banks in the convection path are modeled as porous media that allow prediction of gas-flue resistance, gas temperature, and heat flux because of the complex geometry inside the boiler. Therefore, the detailed ash particle trajectories are not provided. The CFD model obtains the spatial distribution of ash particles to predict fouling deposit information. This information, along with gas velocity and temperature, is fed to the AshPro tool to predict high and low temperature fouling. This method is developed and applied in a tangentially fired 512 MW boiler. The results show the deposit thickness prediction from AshPro on the furnace near the wall and a picture of slagging accumulated at a similar location taken from the boiler are reasonably consistent.

CFD technology is also applied to analyze the erosion occurring on the surface of the heat-transfer boiler of a power plant [111]. The flue gas through the heat-transferring equipment is solved using a finite volume method. The turbulent flow has been solved using Reynolds time averaging for fluctuating components. The governing equation of mass and momentum for a steady state and incompressible flow are defined using the Naiver–Stokes equations. Turbulence is modeled using the standard k-ε turbulence model. The results show that the CFD-based erosion model can match the physical observations. The positions at which larger particles are predicted to hit the surface of heat-transfer equipment are found to correlate with the observed wear. The CFD model is customized to determine the erosion rate of all particles that hit the surface of heat-transfer equipment.

A numerical modeling of coal combustion in a tangentially fired furnace has been created based on CFD technology [112]. The detailed geometry for tubes in the convective passes has not been included in the current CFD model, since the main focus of the current study is on the coal combustion and heat transfer in the radiant section of the furnace. However, source terms for the regions where convective tube banks are located are added to the momentum and energy equations. The heat absorption in the convective tube banks is also considered by adding one of the source terms, which is based on reading data from the process instrument. The coal-combustion process is modeled by chemical reaction. The temperature, composition, and velocity of coal particles along their trajectories are predicted using a Lagrangian particle tracking model. The trajectory of a discrete particle is defined by integrating the force balance on particles in the Lagrangian model. Appropriate forces, such as the drag, gravity, and turbulent dispersion forces, are also considered in the equation of motion in the method.

The balance of mass, momentum, and energy equations in CFD is discretized using the finite volume approach. Two turbulence models of the standard k-ε model and the shear stress transport model are applied to model gas phase turbulence. Both turbulence models provide similar predictions that are in good agreement with the plant data [112].

CFD calculations are used for the calculation of internal and external fluid flows as well as the corresponding pressures, forces, and moments on the heat-transfer surface. Traditionally, the CFD-based calculations run on machines with high-performance computing such as vector machines and reduced instruction set computing machines. An effective parallel-partitioning strategy for an important CFD code is developed using OVERFLOW CFD [113]. OVERFLOW CFD codes are applied to solve viscous compressible flow-related Reynolds-averaged Naiver–Stocks equations with turbulence models. All versions of OVERFLOW are tested and show that it is able to offer promising approaches for taking advantage of parallel hardware.

CFD-based boiler-combustion process simulation is widely applied in the power-generation industry, such as in boiler performance assessment, boiler design, and equipment upgrades. In addition, CFD technology is applied to simulate coal-fired boiler combustion and has achieved satisfactory results [18–20]. Although CFD technology is widely applied in coal-fired boiler combustion, the literature review shows that very few research works focus on a control strategy supported by CFD to limit slagging deposition, which heavily impairs heat-transfer efficiency.

7.5 Optimization Technologies Applied in the Power-Generation Industry

The coal-fired boiler-combustion process is highly complex and multivariable, and uncertain with time delay and time variation, and improving boiler efficiency is a multiobject optimization problem in which compromises are involved between conflicting objectives. Some efforts have been directed toward the developing of a multiobjective-optimization algorithm. A multiobjective Tabu Search (TS) algorithm for continuous-optimization problems is proposed [114]. Two TS algorithms for use on continuous multiobjective-optimization problems are presented. Both algorithms are tested and the results are compared with those from the leading multiobjective genetic algorithm (GA), Non-Dominated Sorting Genetic Algorithm-II (NSGA-II), and the two algorithms perform comparably with NSGA-II.

Multiagent-based software technology has been considered an important approach for developing highly complex industry distributed systems to control and optimize these systems efficiently. Agent-based applications are developed in intelligent manufacturing [115]. Agent encapsulation, agent organization, agent coordination, and negotiation are the key issues in implementing agent-based systems. The function-decomposition and physical-decomposition approaches are contained in the agent encapsulation. In the function-decomposition approach, agents are used to encapsulate functional modules such as order acquisition, process planning, scheduling, materials handling, transportation management, and product distribution. The Hierarchical, Federation, and Autonomous approaches are included in agent organization. The Hierarchical approach is criticized in some literature although it is applied in manufacturing systems. The Federation approach is able to coordinate multiagent activity via facilitation as a means of reducing overhead, ensuring stability, and providing stability. The Autonomous agent approach is suitable for application in distributed intelligent systems. Coordination is a central approach to successful multiagent-based systems, which are highly complex and whose stability is significant. With the ability of an agent to learn from each subsystem, the multiagent-based distributed system can deal with a variable environment and the future performance of the total system will be improved. Therefore, learning is one of the key techniques for a multiagent system.

7.6 Differential Equation–Based Heat-Transfer Process Simulation for a Coal-Fired Power Plant

The differential equations of the coal-fired power plant model have been developed [51]. The model is simplified by utilizing only time derivatives of variables and not spatial derivatives. The superheated steam and furnace exhaust gases are treated as ideal gases. The model is supported by the basic physical thermal dynamics balances as follows.

Heat balance for the superheater, reheater, tubes of the water wall, and economizer in the model is given as

$$Q_{in} + w_{in}h_{in} = w_{ou}h_{ou} + V\frac{\mathrm{d}}{\mathrm{d}t}(\rho h_{ou}) \tag{7.1}$$

where Q_{in} is the incoming heat [J/s], w_{in} is the inlet mass flow [kg/s], h_{in} is the inlet specific enthalpy [J/kg], w_{ou} is the outlet mass flow [kg/s], h_{ou} is the outlet specific enthalpy [J/kg], ρ is the specific density [kg/m³], and V is the volume [m³]. The mass balance in the model is given as

$$w_{in} - w_{ou} = \frac{\mathrm{d}}{\mathrm{d}t}(\rho V) \tag{7.2}$$

The variable definitions in the equation are the same as these of Equation 7.1. The equation of heat radiation in the model is from the Stefan–Boltzmann law:

$$Q = K\theta w_g T_g^4 \frac{1}{\rho_g} \tag{7.3}$$

Q is the heat low from combustion flame radiation [J/s], K is a coefficient $K = 0.18$, θ is the specific angle [rad], w_g is the flow of substances entering combustion [kg/s], T_g is the temperature of the flue gas [K], and ρ_g is the density of combustion flue gas [kg/m³].

The equations of heat transfer due to convection in the model are from engineering experiments. The equation for the heat transfer from combustion gas to the surface of metal tubes is given as

$$Q = K w_g^{0.6} (T_g - T_m) \tag{7.4}$$

where T_m is the temperature of the surface of metal tubes of heat exchangers such as superheaters and reheaters [K]. The definitions of other variables are the same as those of Equation 7.3. The combustion flue gas is assumed as turbulent gas flow.

The equation for the heat transfer from the surface of metal tubes to the steam is given as

$$Q = K w_s^{0.8} (T_m - T_s) \tag{7.5}$$

where T_s is the temperature of steam [K] and the definitions of other variables are the same as Equation 7.4. The steam is assumed as turbulent steam flow in the model.

Although the model showed satisfactory results, it cannot be used to solve slagging-related boiler-combustion problems because slagging is not considered in the model. Furthermore, the spatial derivatives are also not considered in the model. For example, the model cannot be used to identify the slagging distribution in the furnace of the boiler because slagging is spatially distributed on the heat-transfer surface of the water walls of a boiler.

7.7 Existing Problems for Coal-Fired Power Plants

Although advanced technology such as supercritical boiler technology is applied in coal-fired power plants, slagging and fouling still exist and seriously influence the efficiency of power plants. Soot blowing, which is used to remove slagging and fouling in coal-fired power plants, needs improvement [116]. Because of the nonlinearity and multi-input and multioutput of coal-fired boiler combustion, artificial intelligence (AI)-based optimization methods such as NNs and the genetic algorithm method are applied to optimize coal-fired boiler combustion. However, with fouling and slagging occurring inside the furnace, AI-based optimization methods are not always successful because of the nonavailability of data regarding slagging and fouling to train the NNs, so some research focusing on the mechanism of slagging and fouling was undertaken [11]. Therefore, it is significant to improve boiler efficiency and decrease carbon emissions by optimizing coal-fired boiler combustion.

7.8 Conclusion

This chapter has reviewed and discussed the main methods currently used to improve thermal power plant boiler efficiency. Furthermore, the existing thermal power plant efficiency problems are also discussed near the end of the chapter. New methods for solving existing thermal power plant efficiency problems are discussed in Chapters 8 through 10.

8

Online Learning Integrated with CFD to Control Temperature in Combustion

8.1 Introduction

Proportional–integral–derivative (PID)-based control systems are widely applied in the electrical power generation industry; however, for highly complex combustion problems, such as slagging and fouling, conventional PID controllers have limitations. It is very difficult to use PID controllers to control combustion processes and prevent slagging and fouling because data regarding slagging and fouling are very difficult to measure, and a PID controller cannot work on a control system in which sufficient output parameters are not available. Slagging and fouling badly influence boiler-combustion efficiency [3,11], so it is significant to solve the problems for coal-fired power plants with high-slagging trends in the combustion process. This chapter proposes a novel control method of integrating online learning with computational fluid dynamics (CFD) to control the temperature in a generic combustion process, and discussion comparing the method with a PID controller is included.

8.2 Boiler-Combustion Process

Artificial intelligence (AI)-based methods have been applied in a number of power stations to optimize the combustion process and increase efficiency [5–9]. In addition, CFD is used to model the complex combustion process, which has achieved great success for assessment of boiler performance [10–15]. This section presents the boiler-combustion process of a coal-fired power plant.

Figure 8.1 shows the heat-flux distribution inside the furnace of a boiler in which fuel with specific characteristics is sent to the mill, where the coal is pulverized and blown into the furnace of the boiler from burners by mixing with the primary air. Measurement point 1 can measure the amount of

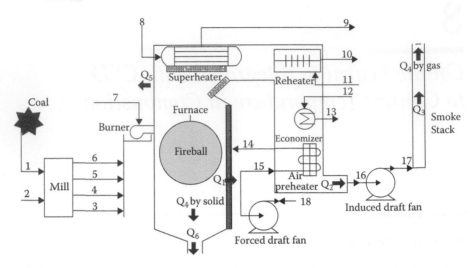

FIGURE 8.1
Heat-flux distribution in the boiler-combustion process.

coal which is sent to mill and point 2 can measure the amount of primary air which is mixed with coal powder. The pulverized coal concentration can be measured at point 3. The speed of the flow of the mixture of air and coal powder can be measured at point 4. The temperature of the mixture of coal and air can be measured at point 5. The excess air rate can be measured at point 6. All the boiler input parameters can be adjusted and tuned in the control system, such as a programmable logic controller (PLC)-based control loop or distributed control system (DCS) in a power plant.

The burners are installed in the water walls of the furnace and the mixture of fuel and air are blown into the furnace to burn. The secondary air is applied to adjust the flame shape of the fireball. It can be measured from point 7. With appropriate adjustment of the angle of the burners, a rotary fireball can be formed in the furnace of the boiler. A high percentage of heat radiates to the surface of the water wall and the superheaters. At the same time, heat conduction occurs on the heat-transfer surface of all equipment inside the furnace. With flue gas flowing in the convection path of the furnace, the residual heat is transferred outside the boiler. The saturated steam, which is heated in the boiler, drives the turbine with high enthalpy due to its high temperature and pressure. The measurement points 8 and 9 can measure the characteristics of the saturated steam in the primary and secondary superheaters. The reheater, economizer, and air preheater are installed in the gas-flue path to recover the residual heat. The temperature and pressure of the steam in the reheater can be measured from points 10 and 11. The temperature of feed water in the economizer is measured from points 12 and 13. The temperature of the primary air inside the reheater can be measured from points 14 and 15.

The residual flue gas blows out from the smoke stack and the temperature and pressure of the exit gas can be measured from point 16. A forced draft (FD) fan and an induced draft (ID) fan keep the correct draft inside the furnace. The power of the fans can be measured from points 17 and 18.

This is a simple normal-combustion process occurring inside the furnace of a boiler. However, the real combustion process is much more complex than this simple process. A number of chemical reactions and physical activities, such as slagging deposition, corrosion, and erosion, occur on the heat-transfer surfaces, which impair the efficiency of combustion. Therefore, lots of energy is lost and global warming gases are emitted.

How and where is the heat lost and emissions increased inside the boiler? Equation 5.18 gives an expression of the heat balance in the combustion process of the boiler [70].

As clarified in Chapter 5, Q_1 is the heat absorbed by the water and steam inside the tubes of the water wall, superheater, reheater, and economizer. It also includes the heat recovered from the preheater, where the cold air absorbs the heat of the residual flue gas. Normally, Q_1 is in the range of 75%–90%. However, slagging, which accumulates on the heat-transfer surface, can severely influence Q_1. More slagging on the surface of the water wall, superheater, and other heat-transfer equipment can massively decrease the heat radiation by which the heat radiates from the flame of the fireball to the heat-transfer surface. At the same time, more slagging can cause less heat conductivity in which the heat is less massively and rapidly transferred to the water or steam side than a system with less slagging on the surface of heat-transfer equipment.

In addition, more slagging can increase the blockage in the flue-gas pass and decrease the convection in which heat can convect rapidly to other heat-transfer equipment from the surface of the fireball. Moreover, more blockage caused by the slagging in the flue-gas pass can increase the power consumption of the FD and ID fans, which decreases the overall efficiency of power plant. Slagging is built up on the furnace walls, which are mainly in the radiation section. It is in a highly viscous state and forms a liquid layer. Fouling is built up by condensed materials. It is a dry deposit and generally in the convection section.

Q_2 is the heat carried by the exit flue gas, which includes the water vapor, oxygen, nitrogen, carbon dioxide (CO_2), and other gas transporting the residual heat to the atmosphere. The higher temperature and higher exit gas volume can cause more residual heat loss from the furnace of the boiler. On the other hand, too low a temperature of exit gas can cause more chemical corrosion on the surface of the heat-transfer equipment installed in the flue-gas path near the exit. The two outputs of heat loss and corrosion conflict with each other with some of the same parameters such as exit gas temperature and pressure.

Q_3 is the heat contained in the combustible gas, such as CO, H_2, and CH_4, which are unburnt and emit with exit flue gas.

Q_4 is the heat contained in the carbon that is unburnt and lost with clinkers dropped outside of the furnace. The proposed solution in this work tries to maintain the powder cloud of the mixture of coal and air a little longer in the exact position of the furnace and maintains the optimum temperature of the fireball by controlling the speed of the rotating fireball, the amount of fuel flow in the pipe, and the tilt angle of the burners.

Q_5 is the heat loss outside of the furnace of the boiler. Normally, it can be decreased by improving the insulation condition of the outside surface of the furnace and is much less than Q_2, Q_3, and Q_4, respectively. Q_6 is the heat carried by the clinkers that are dropped to the outside of furnace from the furnace bottom and is much less than Q_2, Q_3, and Q_4, respectively.

The data acquisition system and DCS, which primarily apply a PID-control strategy based on input, state, and output variables in a measurable process, are widely applied in the power generation industry [70]. However, there are also some immeasurable processes in which a number of critical parameters are very difficult to measure. For example, some parameters including slag thickness, slag accumulation, and corrosion rate are difficult to read using traditional instruments from the boiler-combustion process, which is highly complex and significant process in a power plant because of the states of all existing equipment, flue-gas behavior, and work fluid physical or chemical properties.

8.3 Integrating Online-Learning Technology with CFD-Based Real-Time Simulation to Control the Combustion Process

This work aims to integrate an online-learning controller with an online-simulation module to control a complex combustion process, in which there are some critical process variables that are not easy to measure using industry instruments.

First, a neural network–based adaptive controller with the ability to learning a real-time process is developed. This work consists of designing an online indirect adaptive controller based on a radial basis function (RBF) combined with a numerical-combustion process, which is simulated using CFD. Second, the integrated system is simulated in Simulink. Finally, another PID controller is built, which substitutes the proposed online-learning controller combined with a CFD-based simulation module to validate the proposed control system. The performance of the two different controllers is compared and the results show that the online-learning controller is more efficient than the PID controller. Moreover, the work shows encouraging results in which integrating an online-learning controller with a CFD-based online-simulation module can provide a new strategy to control a complex combustion process in which reading instrument data is difficult.

8.3.1 Online-Learning Technology Method

An RBF neural network–based indirect adaptive-control methodology is applied to build an online-learning model and gradient descent method to improve RBF neural network weights. The control of plants can be described as

$$y(k+d) = f(x(k), u(k)) \tag{8.1}$$

where $f(x(k), u(k))$ is a function of its input variables $u(k)$ and state variables $x(k)$, $y(k)$ is the output, and d is the delay between input and output. The RBF indirect adaptive-control method is given as [117,118]

$$y(k+d) = \alpha(x(k)) + \beta(x(k))u(k) \tag{8.2}$$

The reference input $r(k + d)$ is specified by the user. In an "ideal" controller, $\alpha(x(k))$ and $\beta(x(k))$ are unknown functions, so

$$u(k) = \frac{-\alpha(x(k)) + r(k+d)}{\beta(x(k))} \tag{8.3}$$

$$u(k) = \frac{-\hat{\alpha}(x(k)) + r(k+d)}{\hat{\beta}(x(k))} \tag{8.4}$$

$$\hat{\alpha}(x(k)) = \theta_\alpha^T(k)\phi_\alpha(x(k)) + \alpha_k(x(k)) \tag{8.5}$$

$$\hat{\beta}(x(k)) = \theta_\beta^T(k)\phi_\beta(x(k)) + \beta_k(x(k)) \tag{8.6}$$

A normalized gradient method is used to train $\theta(k)$ to approximate $\alpha(x(k))$ and $\beta(x(k))$, which is given as

$$\theta(k) = \theta(k-d) + \lambda_k d(k) \tag{8.7}$$

where λ_k is the step size and $d(k)$ is the descent normalized gradient.

$$\theta(k) = \theta(k-d) + \frac{k_1 \eta \theta(k-d)}{1 + \gamma |\theta(k-d)|^2} e_\varepsilon(k) \tag{8.8}$$

where $k_1 = 1$ and γ is a design parameter.

$$e_\varepsilon(k) = \begin{cases} e(k) - \varepsilon(k) & \text{if } e(k) > \varepsilon(k) \\ 0 & \text{if } |e(k)| \leq \varepsilon(k) \\ e(k) + \varepsilon(k) & \text{if } e(k) < \varepsilon(k) \end{cases} \tag{8.9}$$

where $e(k)$ is a track error modification of which $e_\varepsilon(k).\varepsilon(k) = t_1 + t_2 \mid u(k-1) \mid$. In this work, tuning parameters t_1 and t_2 are considered to be 0.01.
Suppose

$$\hat{\alpha}(y(k)) = \theta_\alpha^T \phi_\alpha(y(k)) \tag{8.10}$$

$$\hat{\beta}(y(k)) = \theta_\beta^T \phi_\beta(y(k)) \tag{8.11}$$

where θ_α and θ_β are the parameter vectors and ϕ_α and ϕ_β represent the RBF neural network defined below.

The RBF network mapping is of the following form [117,118]:

$$F_{rbf}(x, \theta_{rbf}) = \sum_{i=1}^{n} b_i R_i(x) \tag{8.12}$$

where θ_{rbf} is θ_α or θ_β.

$$R_i(x) = \exp(-\frac{|x-c^i|^2}{(\sigma^i)^2}) \tag{8.13}$$

where x is the input vector, and c^i and σ^i are parameters that enter in a non-linear fashion.

Let $\theta_{rbf} = [b_1, b_2, \ldots, b_n]^T$ and $\phi_{rbf} = [R_1, R_2, \ldots, R_n]^T$, then

$$F_{rbf}(x, \theta_{rbf}) = \theta_{rbf}^T \phi_{rbf} \tag{8.14}$$

From Equations 8.3, 8.4, 8.6, 8.7, and 8.9, we obtain the online-learning controller input

$$u(k) = \frac{-\hat{\alpha}(y(k)) + r(k+d)}{\hat{\beta}(y(k))} \tag{8.15}$$

where $r(k)$ is the command input and $y(k)$ is the output of the online-simulation plant process.

8.3.2　CFD Model Method

CFD is concerned with the numerical solution of differential equations governing transport of mass, momentum, and energy in thermal dynamics, fluid dynamics, and chemistry dynamics. The equations are applied to simulate the combustion process and can be written as follows. A control volume (CV) is defined as fixed in space and the fluid is assumed to flow through the CV.

A CV is assumed to be located at (x_1, x_2, x_3) [71]. The momentum equation is given as

$$\frac{\partial(\rho_m u_i)}{\partial t} + \frac{\partial(\rho_m u_j u_i)}{\partial x_j} = \frac{\partial}{\partial x_j}\left[\mu_{eff}\frac{\partial u_i}{\partial x_j}\right] - \frac{\partial p}{\partial x_i} + \rho_m B_i + S_{u_i} \qquad (8.16)$$

where ρ_m is the density of the fluid (kg/m³), x_i, x_j are displacements (*m*) in the *i* and *j* directions, u_i, u_j are velocities in the *i* and *j* directions, μ_{eff} is the viscosity of the fluid, B_i is the body force (N/kg), and S_{u_i} is the other source of momentum.

The equation of mass transfer can be written as

$$\frac{\partial(\rho_m w_k)}{\partial t} + \frac{\partial(\rho_m u_j w_k)}{\partial x_j} = \frac{\partial}{\partial x_j}\left[\rho_m D\frac{\partial w_k}{\partial x_j}\right] + R_k \qquad (8.17)$$

where w_k is the mass fraction, $w_k = \frac{\rho_k}{\rho_m}$, and $\sum_{\text{all species}} w_k = 1$. D is the mass diffusivity and R_k is the rate of generation in the CV. The energy equation is given as

$$E = Q_{\text{conv}} + Q_{\text{cond}} + Q_{\text{gen}} - W_s - W_b \qquad (8.18)$$

where E is the rate of change of energy of the CV, Q_{conv} is the net rate of energy transferred by convection, Q_{cond} is the net rate of energy transferred by conduction, Q_{gen} is the net volumetric heat generation within the CV, W_s is the net rate of work done by surface forces, and W_b is the net rate of work done by body forces.

A general combustion model is used to show how to combine an online-learning controller with a CFD model in the presented work.

The dynamic system consists of a metal block that exchanges heat with the environment. A heater, which is controlled by the online-learning controller, is situated inside the glass-enclosed system. The system works as follows. The heater is controlled by the proposed controller and a specific constant temperature is kept at any chosen point inside the metal block. Figure 8.2 shows the geometry model built in Comsol 3.2, which is a CFD-based commercial software product.

A point is selected that is required to be controlled by the proposed learning controller. Figure 8.3 shows the temperature-field distribution at the fifth second. This shows that the temperature is gradually dropping from the heat source point to each side. Figure 8.4 shows the temperature distribution at the selected point in 5 seconds and the temperature reaches the maximum at the fifth second. Figure 8.5 shows the temperature-field distribution within 5 seconds at different points that are on the same vertical line as the selected point. It shows that the closer the point to the heat source center, the higher the temperature it can reach at the same time.

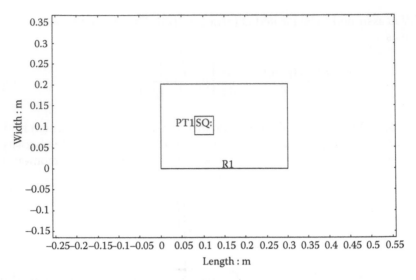

FIGURE 8.2
A simple combustion model in Comsol Multiphysics 3.2.

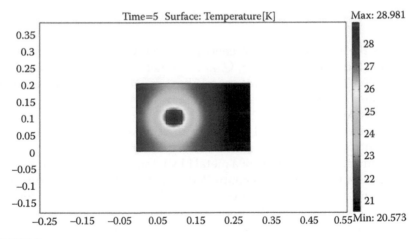

FIGURE 8.3
Temperature-field distribution of the combustion process at the fifth second.

8.3.3 Integrating Online Learning with CFD

The strategy of the proposed control methodology is shown in Figure 8.6. Compared with the traditional indirect adaptive control, the online-simulation module is added in the control loop. The online-learning controller is trained using real-time data, which flows from the online-simulation module rather than the plant processes. In some complex plant processes,

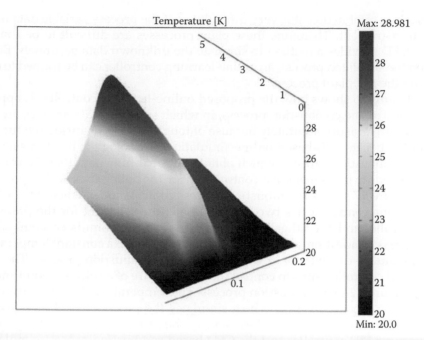

FIGURE 8.4
Temperature-field distribution of the combustion process from 0 to 5 s.

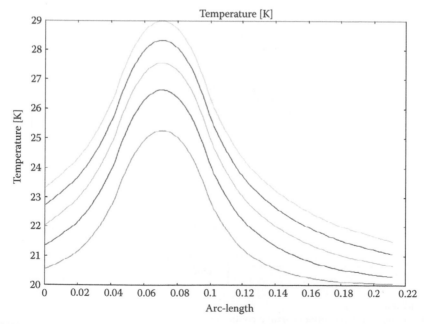

FIGURE 8.5
Temperature-field distribution of the combustion process from 0 to 5 s.

such as combustion, it is very difficult to get the process variable data using an instrument. Therefore, these plant processes are difficult to be control. A CFD provides a method to simulate the unknown data accurately. Based on the simulated process, an online-learning controller can be trained to control the real plant process.

Figure 8.6 shows how the proposed online-learning controller is applied to control the combustion process, in which some critical variables are difficult to measure accurately because of tough conditions inside the furnace of the boiler. CFD-based online-simulation technology is used to simulate the combustion process, which obtains the real-time data that the controller needs to implement in a control loop for the boiler-combustion process.

Figure 8.7 shows a combustion process control model, where the online-learning controller has two inputs. One is the set point for the proposed controller and the other is the feedback from the simulated combustion process. In addition, a white noise signal is added to a constant temperature to simulate the exterior environment of the combustion process. The proposed controller aims to control the temperature of a selected point shown in Figure 8.2 in a combustion process. The temperature data from the combustion process module, which simulates the process based on CFD, are fed to the input parameter heat state of the proposed controller. Figure 8.8 shows a PID controller and the CFD-based combustion-process simulation which are integrated to evaluate the system. The performance of the two controllers in Figures 8.7 and 8.8 is compared and discussed in Section 8.4.

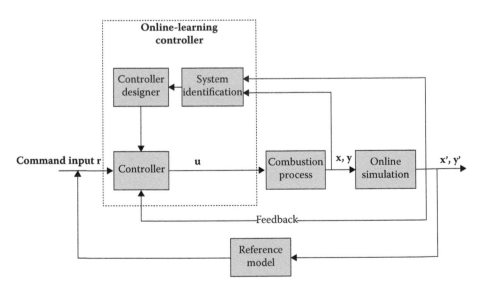

FIGURE 8.6
Logic of the proposed online-learning controller.

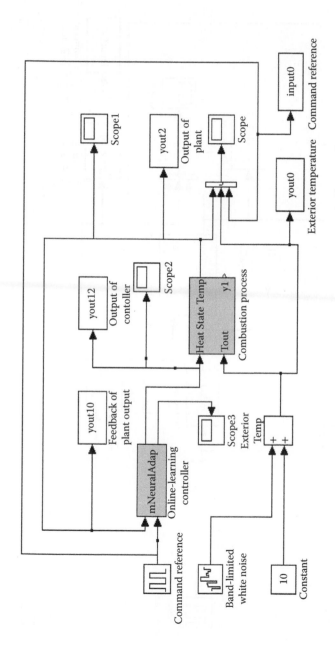

FIGURE 8.7
Integrating online-learning controller with combustion process CFD model in Simulink.

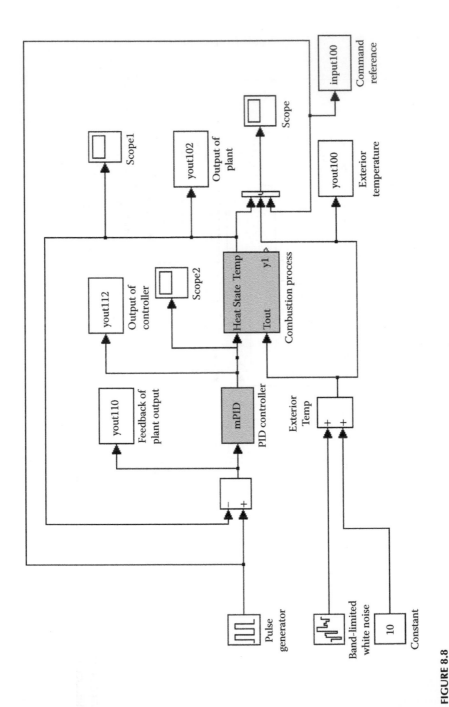

FIGURE 8.8
Integrating PID controller with combustion process CFD model in Simulink.

8.4 Results and Discussion

A constant and a square wave are the inputs to the online-learning controller to evaluate the system. Figure 8.9 shows that the online-learning controller (neural network controller) tracks the control object quicker than a PID controller and has less oscillation.

Figures 8.10 and 8.11 show that the temperature of a selected point in combustion space is better controlled and stable under the proposed online-learning controller than with a PID controller, and the temperature of a selected point becomes stable faster using an online-learning controller rather than a PID controller. In a real combustion process, if the load of a boiler is required to change frequently, that also means the fuel control has to change frequently to meet the load requirement. Therefore, in these conditions, an online-learning controller can consume less fuel than a PID controller.

Figure 8.12 shows that the online-learning controller controls more exactly than the PID controller with a constant command reference. Figure 8.13 shows that the online-learning controller is more stable than the PID controller with a square wave that is used as a reference input for the controllers, although both controllers have oscillations. Figures 8.14 and 8.15 show that the online-learning controller tracks the control object in a quicker and more stable manner than the PID controller under a square wave, which is used as a reference input. Figure 8.16 shows that the online-learning controller tracks the control object in a more accurate manner than the PID controller under a square wave which is used as a reference input.

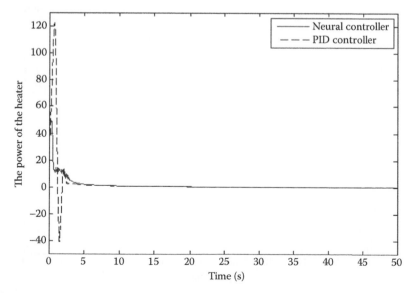

FIGURE 8.9
Outputs of the online-learning controller and PID controller.

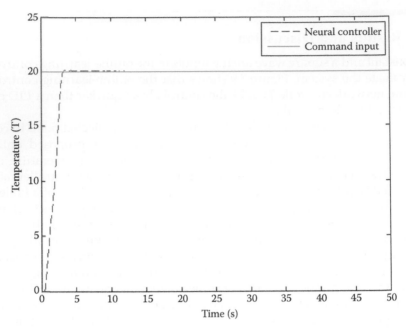

FIGURE 8.10
Temperature of a point controlled in the combustion process with a neural network controller.

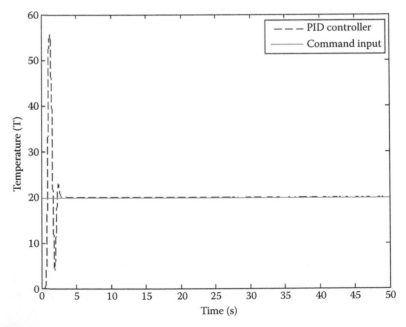

FIGURE 8.11
Temperature of a point controlled in the combustion process with a PID controller.

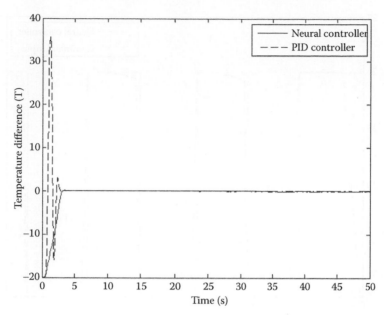

FIGURE 8.12
Comparison of the error of the system controlled respectively by the two controllers.

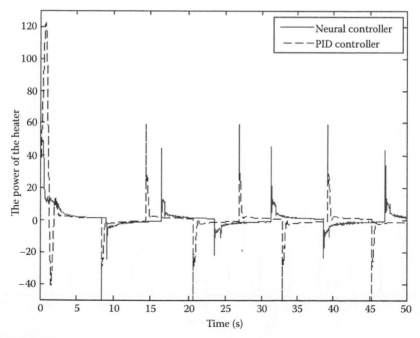

FIGURE 8.13
Output of the controller while using a square wave as a reference input of the controller.

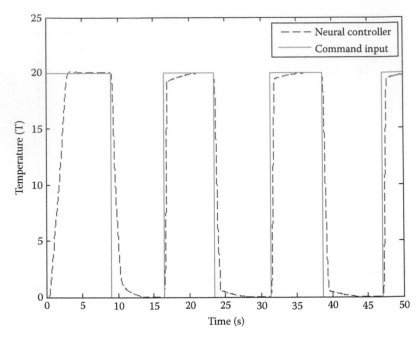

FIGURE 8.14
Plant output while using a square wave as a reference input of the neural network controller.

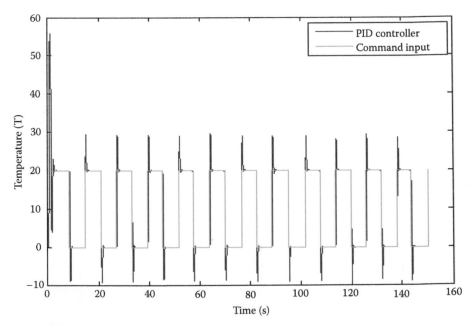

FIGURE 8.15
Plant output while using a square wave as the reference input of a PID controller.

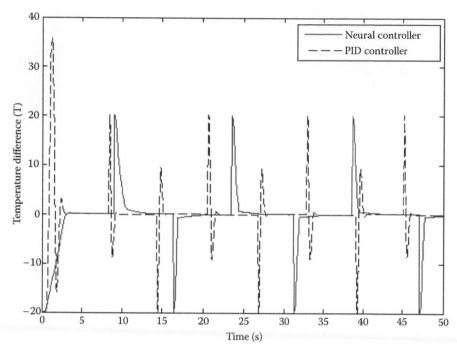

FIGURE 8.16
Error of the system while a square wave is used as the reference input of the controller.

8.5 Conclusion

This chapter has developed a novel method of integrating online learning with CFD to control the combustion in which the temperature data in the combustion is obtained by a CFD-based combustion model rather than instrument readings. The comparison between the proposed method and the PID-based control is also discussed in the chapter, and the results show that control integrating online learning with CFD can achieve superior performance. Chapter 9 discusses how to integrate a computational intelligence method with CFD modeling to improve thermal power plant boiler efficiency.

9

Online Learning Integrated with CFD to Identify Slagging and Fouling Distribution

9.1 Introduction

Soot blowing is designed to remove slagging and fouling and is frequently used by an operator on each shift in a power plant. Slagging- and fouling-related combustion problems still exist and have a severe influence on a power plant in which coal has a higher tendency of slagging and fouling, so how to apply soot blowers intelligently rather than frequently to clean the heat-transfer surface and maintain a high heat-transfer efficiency inside the furnace is indispensable. In addition, intelligent soot blowing can not only reduce steam loss from unexpected blowing and avoid tube damage caused by overblowing [119], but also monitor the slagging by quantifying the slagging and its distibution. This is significant to a coal-fired power plant's safe and economical operation.

Artificial intelligence (AI) is often used in a power plant to optimize performance and tune conventional proportional–integral–derivative (PID)-based control because of the complexity and nonlinearity of the model for power plant processes such as the combustion process [120,121]. A recurrent neural network (NN) is applied to model a large scale once-through type of ultrasupercritical boiler of a power plant. The NN-based predictive model can drive the plant to a desired state by tuning the PID-based control system based on the optimal values of gains that are obtained from the model. AI-based methods are also applied in coal-fired power plants of the US to optimize performance. For example, the NN-based method has been applied to make the conventional soot-blowing system operate with more intelligence, which improved heat-transfer efficiency and lowered emission of nitrogen oxides (NO_x) [122]. A neural network technique is used to improve the control system by obtaining optimal parameters in a predictive fashion [9,82,83,123,124]. However, for combustion-related problems, the technology, which is only dependent on AI, does not work successfully because not many readings and data regarding the combustion process can be obtained to train the NN-based models and acquire approximate functions for such complex processes.

On the other hand, computational fluid dynamics (CFD) is a part of simulation technology, which is applied in industries for thermal design and analysis [125]. It has been successfully applied to simulate highly complex industry processes [67,68,126,127]. CFD is also used to model and analyze the combustion process for a power plant boiler [86,94]. For example, modeling using CFD has been developed to predict the air and pulverized coal behavior and get an exact optimal boiler geometry size for a boiler design [95–99]. A CFD-based model is also used to adjust burner-tilt angle offline after overhaul or upgrade in a power plant [100].

In fact, with advancement of computer technology and mathematical methodology, integrating AI with CFD technologies can solve combustion-related problems. Based on this, the research proposes new methods to improve combustion-process efficiency and decrease carbon emissions for the fossil fuel power-generation industry. The new methods can control, identify, and optimize the coal-fired power plant boiler-combustion process by integrating online learning with CFD technology. This chapter focuses on integrating online learning with CFD technology to identify slagging and fouling distribution inside the furnace of a coal-fired power plant boiler.

9.2 Multiobjective Online Learning

9.2.1 The Proposed Multiobjective Learning System

Online-learning-based real-time control is applied in a number of industry areas. However, many of them are used in the fashion in which the NN has only one output. This research creates an NN-supported method that can identify multiple objectives in a process. Consider the control of a nonlinear system [117,128]:

$$x_1(k+1) = x_2(k)$$

$$\ldots \tag{9.1}$$

$$x_n(k+1) = f(x(k)) + u(k)$$

where $x(k) = [x_1(k), x_2(k), \cdots, x_n(k)]^T \in R^{nm}$ with each $x_i(k) \in R^m$, $i = 1, \ldots, n$ is the state at time instant k, $f(x(k)) \in R^m$ is an unknown nonlinear dynamic system, and $u(k) \in R^m$ is the input. The principles for weight update in the NN-based adaptive system are given below [128]. Assuming a trajectory, $x_{nd}(k) \in R^m$, where the trajectory error is

$$e_i(k) = x_i(k) - x_{nd}(k+i-n) \tag{9.2}$$

and the filtered trajectory error is

$$r(k+1) = f(x(k)) - x_{nd}(k+i-n) + \lambda_1 e_n(k) + \cdots + \lambda_{n-1} e_2(k) + u(k)$$

where $e(k) = [e_1(k), e_2(k), \cdots, e_n(k)]^T$. From Equation 4.3, define the input

$$u(k) = x_{nd}(k+1) - \hat{f}(x(k)) + l_v r(k) - \lambda_1 e_n(k) - \cdots - \lambda_{n-1} e_2(k)$$

where $\hat{f}(x(k)) \in R^m$ is an estimate of unknown function $f(x(k))$ and l_v is a diagonal gain matrix. Then

$$r(k+1) = l_v r(k) - \tilde{f}(x(k))$$

where the functional estimation error is

$$\tilde{f}(x(k)) = \hat{f}(x(k)) - f(x(k))$$

Equation 4.5 can also be expressed as

$$r(k+1) = l_v r(k) + \delta_0(k)$$

where $\delta_0(k) = -\tilde{f}(x(k))$; if the functional estimation $\tilde{f}(x(k))$ is bounded as $\left\| \tilde{f}(x(k)) \right\| \le f_M$, and $f_M \in R$, then the next stability result holds. The of the NN is used to approximate $f(x(k))$ and to provide an optimal signal to be a part of $u(k)$:

$$\hat{f}(k) = \hat{w}_2^T(k) \mathbb{0}_2(v_2^T x(k)) = \hat{w}_2^T(k) \mathbb{0}_2(k)$$

where $\hat{w}_2 \in R^{n_2 \times m}$ and $v_2 \in R^{nm \times n_2}$ represent the matrix of weight esti $\mathbb{0}_2(k) \in R^{n_2}$ is the activation function vector in the hidden layer, n_2 number of nodes in the layer, and $x(k)$ is the input of the NN. Suppo unknown target output layer weight of the network is w_2; then we hav

$$f(k) = w_2^T \mathbb{0}_2(v_2^T x(k)) + \varepsilon_2(x(k)) = w_2^T \mathbb{0}_2(k) + \varepsilon_2(x(k))$$

where $\varepsilon_2(x(k)) \in R^m$ is the functional estimation error. Then

$$\tilde{f}(x(k)) = \hat{f}(k) - f(k) = (\hat{w}_2(k) - w_2(k))^T \mathbb{0}_2(k) - \varepsilon_2(x(k))$$

The weight update role is defined as

$$\hat{w}_2(k+1) = \hat{w}_2(k) + \Delta\hat{w}_2(k)$$

The NN is also gradient-based adaption [128]:

$$\Delta\hat{w}_2(k) = -\alpha_2 \mathbb{0}_2(k)(\hat{w}_2^T(k) \mathbb{0}_2(k) + \tilde{f}_2(x(k))^T$$

where $\alpha_2 \in R$ is the adaption gain of the NN.

Substituting Equation 9.12 in 9.11:

$$\hat{w}_2(k+1) = \hat{w}_2(k) - \alpha_2 0_2(k)(\hat{w}_2^T(k)0_2(k) + \tilde{f}(x(k))^T + \tilde{f}(x(k)))$$

From Equation 9.5:

$$\tilde{f}(x(k)) = l_v r(k) - r(k+1)$$

From Equations 9.13 and 9.14:

$$\hat{w}_2(k+1) = \hat{w}_2(k) + \alpha_2 0_2(k)(\hat{w}_2^T(k)0_2(k) + l_v r(k) - r(k+1))^T$$

The principles of the weight-matrix update in Equation 9.15 are us
proposed multiobjective online-learning system method to update
Figure 9.1 shows the logic of the proposed method. The NN car
multiple inputs u_1-u_n to maintain the plant-control system at the re
y to approach the desired reference value r. It also means that for
ing output of a plant system, the proposed method can provide th
inputs to match the existing result. Moreover, the multiple objectiv
can be explored by the proposed method to match the existing resu

Figure 9.2 shows the structure of the proposed multiobjective online
method. The NN is composed of three hidden layers, and the activa
tion is set in the second hidden layer. In addition, the node in the se
third hidden layer can be adjusted dynamically during approximation
put of NN provides multiple inputs u_1-u_n to the real-time CFD-based

The activation function used in the second hidden layer is
function [129]:

$$f(0) = \frac{1}{1 + \exp(-0)}$$

The outputs of the plant system are fed back to the first hidden
the error between reference and feedback is input into the secon
layer of the NN. The weight $w_{31} - w_{3n1}$ is updated based on Equati
each iteration in the third hidden layer. The outputs u_1-u_n are appr
and entered in the plant system. A scenario is used to validate the
multiobjective online learning in Section 9.2.2.

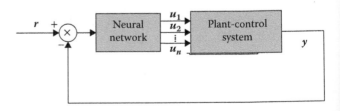

FIGURE 9.1
Logic of the proposed multiobjective online-learning method.

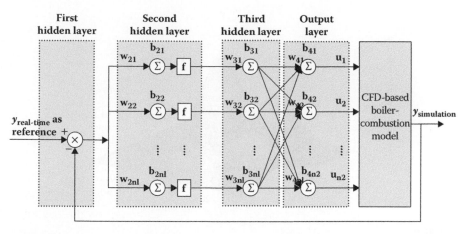

FIGURE 9.2
Structure of the proposed multiobjective online-learning method.

9.2.2 Validation of the Proposed Multiobjective Online Learning

A scenario is set without loss of generality to validate the proposed method. A function for the process in the scenario is denoted in Equation 9.17. This is a nonlinear control process with two inputs u_1 and u_2, and the relationship between the output and the inputs is shown in Figure 9.3.

$$y = 0.8 \times 10^3 (5\exp(-0.005((u_1 - 75)^2 + (u_2 - 70)^2)) + 3\exp(-0.006((u_1 - 85)^2 + (u_2 - 40)^2)) + 2.5\exp(-0.003((u_1 - 35)^2 + (u_2 - 20)^2))) \tag{9.17}$$

The validation results are shown in Table 9.1. In each test case, first, the output of the nonlinear control process is given. Then the multiobjective online method is applied to get the corresponding inputs u_1 and u_2. The relation between the output and inputs u_1 and u_2 can be seen from Figure 9.3. The proposed method can provide two inputs and lead the model output to approach the desired output. The results of the validation are encouraging although they are not exactly equal to the desired values. One of the validation results is shown in Figure 9.4. The result is also highlighted in a contour plot of the plant system in the scenario, which is shown in Figure 9.5. The desired output in this time is set to 3050. Figure 9.5 shows that all pairs of the two input values corresponding to the desired output drop in the specified isoline and the result from the proposed method is specifically marked in the isoline. The proposed method is used to learn inputs for a plant process to meet the expected output, and all results are satisfactory. However, if the proposed method is applied to identify multiple objects, postlearning processing is required to decide on a specific pair to match the desired output, and this is discussed in Section 9.5 of the chapter.

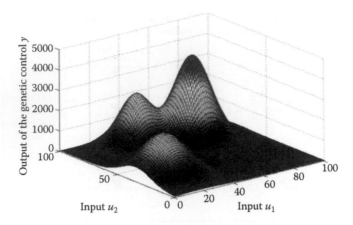

FIGURE 9.3
Relationship between the output and two inputs in the generic-control process.

TABLE 9.1

The Results of the Proposed Method Validation

Test Cases	Desired Output	Model Output	Input u_1	Input u_2
1	3050	3069.8	68.0000	72.0000
2	3050	3136.3	71.0536	64.1225
3	2050	1972.4	63.2103	68.4070
4	1800	1804.7	80.948	47.6300
5	900	977.00	59.6170	76.7290
6	3500	3495.6	75.0374	75.1960

9.3 Modeling of a Power Plant Boiler-Combustion Process Based on CFD

A three-dimensional power plant boiler furnace model is developed using ANSYS Fluent 14.5 based on the real data from a 1160 t/h tangential coal-fired power plant [18]. Flue-gas property fields, such as temperature and intensity, are analyzed and the results show that the simulation output of the flue-gas property is close to the corresponding data from the power-generation industry and simulation results from research [18–20].

9.3.1 Geometry of the Furnace of a Coal-Fired Power Plant

The geometry model is developed based on the data from the power plant. This is a 14.62 m wide, 12.43 m deep, and 48.8 m high furnace. The 44 burners installed are in four corners in a tangential-combustion fashion. The geometry of the furnace is shown in Figure 9.6. The positions of each of the four burners located in a horizontal section are shown in Figure 9.7. The center line of the burner, which is installed in a different corner, is shown in Figure 9.7.

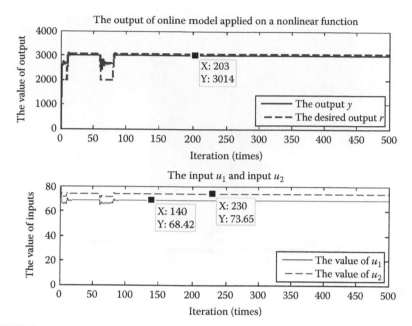

FIGURE 9.4
Identification of the general-control system inputs u_1 and u_2.

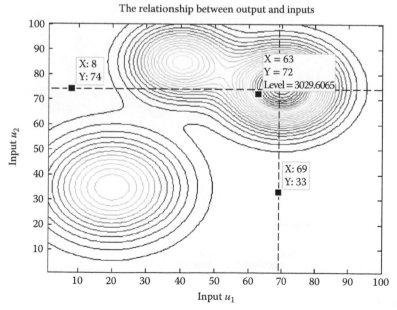

FIGURE 9.5
Contour of the relationship between the output and inputs in the generic-control process.

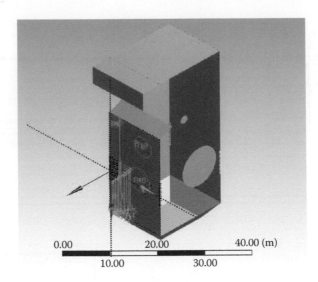

FIGURE 9.6
Geometry of the boiler of a coal-fired power plant developed using ANSYS DesignModeler 14.5.

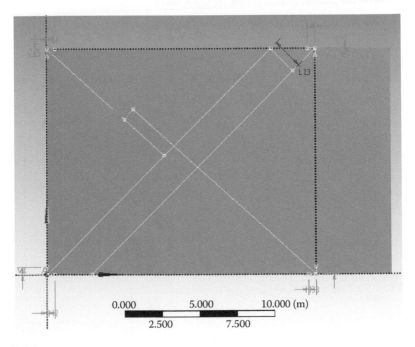

FIGURE 9.7
Center line of each burner viewed from the top of the furnace.

The angle of the center line can decide the diameter of the tangential flame ball and the data are listed in Table 9.2. The pulverized coal property data are shown in Table 9.3 [18], and the burner operating parameters are listed in Table 9.4. The location of each burner group is denoted using values on the z-axis and shown in Table 9.5.

The detailed data of the mesh are listed in Table 9.6. The mesh for the geometry is shown in Figure 9.8. The research combines multiobjective online learning with a CFD-based model to identify the slagging and fouling inside a furnace, and an optimal number for mesh nodes and element is significant, as both insufficient meshing and overmeshing can negatively influence the identification software based on the proposed method.

TABLE 9.2

Characteristics of Tangential Circle of Flame or Gas Fluid

Left of Front to Right of Back of Corner		Right of Front to Left of Back of Corner	
Angle of center line	Distance between two lines	Angle of center line	Distance between two lines
42°	0.89 m	45°	2.12 m

TABLE 9.3

The Property of Pulverized Coal

Proximate Analysis		Ultimate Analysis	
Volatile	0.513	C	0.766944
Fixed carbon	0.472	H	0.032966
Ash	0.015	O	0.194293
Moisture	0	N	0.004348
–	–	S	0.001449
Coal ash received HCV (J/kg)		2.9×10^7	

TABLE 9.4

Operating Parameters of Pulverized Coal Burner

Primary Air		Secondary Air		Boundary Air	
Velocity (m/s)	Temperature (°C)	Velocity (m/s)	Temperature (°C)	Velocity (m/s)	Temperature (°C)
25.4	76	54.11	331	54.11	331

TABLE 9.5

Location of Burners in Each Burner Group

Group Number (Number for Burner in Each Level)	Air Type	Value of z-Axis (m)
1	Secondary air	27.026
2	Primary air	27.628
3	Secondary air	28.179
4	Primary air	28.833
5	Secondary air	29.315
6	Primary air	29.970
7	Secondary air	30.469
8	Primary air	31.1575
9	Secondary air	31.6225
10	Primary air	32.2595
11	Secondary air	32.7665

TABLE 9.6

Data of Mesh for the Geometry

Min Size	7.6958×10^{-3} m
Max Size	1.53920 m
Max Face Size	0.769580 m
Nodes	56,002
Elements	306,763

9.3.2 Modeling the Combustion Process

Heat transfer including conduction, convection, and radiation occurs in different sections of the furnace. In the center of the furnace, heat transfers to the metal surface of the water wall pipes from the flame of burning pulverized coal by radiation. Then the heat transfers to the water side of the metal pipes through conduction and the heat can be absorbed by the flowing water or mixture of steam and water by conduction and convection. In the flue-gas path, heat is carried to the metal surface of the superheater and reheater. Then the heat can be absorbed by convection. Finally the steam inside the pipes of the superheater or reheater can absorb the heat by conduction and convection. The entire process of pulverized coal combustion is modeled in the research.

Equation 9.18 models the conductive and convective heat transfer [19–21]:

$$\frac{\partial x}{\partial t}(\rho E) + \nabla(\overline{v(\rho E + p)}) = \nabla \cdot \left(k_{\text{eff}} \nabla T - \sum_j h_j \overline{j}_j + (\overline{\overline{\tau}}_{\text{eff}} \cdot \overline{v}) + S_h \right) \quad (9.18)$$

where ρ is the intensity, p is the pressure, $\overline{v}$ is the velocity, k_{eff} is the effective conductivity, $\overline{j}_j$ is the diffusion flux of species j, h_j is the enthalpy of species of

0.000   15.000   30.000 (m)

        7.500   22.500

FIGURE 9.8
Mesh of the geometry model using ICEM CFD 14.5.

j, $\overline{\overline{\tau}}_{\text{eff}}$ is the viscosity of the flue, and term S_h is the amount of heat from chemical reaction and any other defined heat source. In Equation 9.18, the energy
transfer due to conduction is defined as

$$q_{\text{cond}} = k_{\text{eff}} \nabla T \tag{9.19}$$

The energy transfer due to species diffusion is defined as

$$q_{\text{diff}} = \sum_j h_j \overline{j}_j \tag{9.20}$$

The energy transfer due to viscous dissipation is defined as

$$q_{\text{diss}} = \overline{\overline{\tau}}_{\text{eff}} \cdot \overline{v} \tag{9.21}$$

In Equation 9.18,

$$E = h - \frac{P}{\rho} + \frac{v^2}{2} \tag{9.22}$$

where sensible enthalpy h is defined for an ideal gas as

$$h = \sum_j Y_j h_j \tag{9.23}$$

In Equation 9.23, Y_j is the mass fraction of species j and

$$h_j = \int_{T_{ref}}^{T} c_{p,j} dT \tag{9.24}$$

where T_{ref} is 298.15 K, and $c_{p,j}$ is the specific heat-capacity rate of species j.

Pulverized coal combustion is a nonadiabatic, nonpremixed process, and the total enthalpy form of the energy in the model is given as

$$\frac{\partial y}{\partial x}(\rho H) + \nabla \cdot (\rho \bar{v} H) = \nabla \cdot \left(\frac{k_t}{c_p} \nabla H \right) + S_h \tag{9.25}$$

where ρ is the intensity, p is the pressure, $\bar{v}$ is the velocity, k_t is the conductivity of flue gas in turbulent combustion, and c_p is the specific heat capacity rate. In Equation 9.25, the total enthalpy H is defined as

$$H = \sum_j Y_j H_j \tag{9.26}$$

where Y_j is the mass fraction of species j and

$$H_j = \int_{T_{ref}}^{T} c_{p,j} dT + h_j^0 (T_{ref,j}) \tag{9.27}$$

$h_j^0 (T_{ref,j})$ is the formation enthalpy of species j at the reference temperature T_{ref}.

In Equation 9.25, the chemical reaction energy source S_h is defined as

$$S_h = -\sum_j \frac{h_j^0}{M_j} R_j \tag{9.28}$$

where h_j^0 is the enthalpy of formation of species j and R_j is the volumetric rate of creation of species j, and M_j is the molecular mass of species j. In the metal pipes of the water wall, superheater, and reheater, the energy equation is given as [21]

$$\frac{\partial}{\partial t}(\rho h) + \nabla \cdot (\bar{v} \rho h) = \nabla \cdot (k \nabla T) + S_h \tag{9.29}$$

where ρ is the density, h is the enthalpy, k is the conductivity, T is the temperature, and S_h is the volumetric heat source.

The radiation-transfer equation for an absorbing, emitting, and scattering medium at position $\bar{r}$ in direction $\bar{s}$ is given as [21–23,50]

$$\frac{dI(\vec{r},\vec{s})}{ds} + (a+\sigma_s)I(\vec{r},\vec{s}) = an^2\frac{\sigma T^4}{\pi} + \frac{\sigma_s}{4\pi}\int_0^{4\pi} I(\vec{r},\vec{s})\emptyset I(\vec{s}\cdot\vec{s}')d\Omega'$$

where $\vec{r}$ is the position vector, $\vec{s}$ is the direction vector, $\vec{s}'$ is the scat
direction vector, s is the path length, a is the absorption coefficient, n
refractive index, σ_s is the scattering coefficient, σ is the Stefan–Boltz
constant ($5.66\times\frac{10^{-5}}{m^2}$ (W/m$^2\cdot$K^4)), I is the radiation intensity, which de
on position $\vec{r}$ and direction $\vec{s}$, T is the local temperature, $\emptyset$ is the phase
tion, and Ω' is the solid angle.

Energy coupling and the discrete ordinates (DO) model [20] are appl
the research to simulate the heat-radiation process inside the furnac
DO model considers Equation 9.30 in direction $\vec{s}$ as the field equation
the equation is written as

$$\nabla\cdot(I(\vec{r},\vec{s})\vec{s}) + (a+\sigma_s)I(\vec{r},\vec{s}) = an^2\frac{\sigma T^4}{\pi} + \frac{\sigma_s}{4\pi}\int_0^{4\pi} I(\vec{r},\vec{s})\emptyset I(\vec{s}\cdot\vec{s}')d\Omega'$$

The energy equation when integrated over a control volume i can g
model of coupling between energy [21–23,50]. The model is present
follows:

$$\sum_{j=1}^{N}\mu_{ij}^T T_j - \beta_i^T T_i = \alpha_i^T\sum_{k=1}^{L} I_i^k\omega_k - S_i^T + S_i^h$$

where $\alpha_i^T = k\Delta V^i$, $\beta_i^T = 16k\sigma T_i^3\Delta V_i$, $S_i^T = 12k\sigma T_i^4\Delta V_i$, k is the absorption
cient, and ΔV is the control volume. The coefficient μ_{ij}^T and the source te
are due to the discretization of the convection and diffusion terms.

The research focuses on optimizing the coal-fire combustion p
in which pulverized coal and oxide air enter the reaction zone in di
streams. Compare with premixed system in which reactants are mi
the molecular level before reaction, pulverized coal combustion is a nc
mixed system, so a nonpremixed-combustion model [18–22] is applied
research. The basis of the model is that the instantaneous thermoche
state of the fluid is related to a conserved scalar quantity known as the
ture fraction, f, which is given as

$$f = \frac{z_i - z_{i,ox}}{z_{i,fuel} - z_{i,ox}}$$

where z_i is the element mass fraction for element i. The subscript ox de
the value at the oxidizer stream inlet and the subscript fuel denote
value at the fuel stream inlet. The transport equation for the mixture fra
[21–23,50] is given as

$$\frac{\partial}{\partial t}(\rho\bar{f}) + \nabla\cdot(\rho\vec{v}\bar{f}) = \nabla\cdot\left(\frac{\mu_t}{\sigma_t}\nabla\bar{f}\right) + S_m$$

where ρ is the density, $\bar{f}$ is the mean mixture fraction, $\bar{v}$ is the local velocity, μ_t is the turbulent viscosity, the constant $\sigma_t = 0.85$, and the source term S_m is solely due to transfer of mass into the gas phase from the pulverized coal particle.

ANSYS Fluent 14.5 is applied in this research to develop a coal-fired power plant boiler-combustion process model. The 44 burners are the inputs and the exhausted gas exit is the output of the model. The data for the boundary conditions are listed in Tables 9.3 and 9.4. The coal-fired combustion process including chemical reactions, the heat-radiation process in the radiation section, the heat-convection process occurring in the flue-gas path, and the heat-conduction process occurring between the fire and water side of the water wall are all simulated in the developed model of the research. Compared with the real data in the power-generation industry and simulation results in other research, the results of the boiler coal-fired power plant boiler-combustion model are encouraging, and they are discussed in the next section.

9.4 Analyzing the Results of the Boiler-Combustion Process Model

The developed coal-fired power plant boiler-combustion model is a three-dimensional model in which the fields of gas temperature, velocity, pressure, and intensity are all simulated. In addition, the radiation and chemical reaction regarding nitrogen and carbon are simulated. The temperature field is first analyzed. Second, the incident radiation is discussed. It is also compared with the temperature distribution and both results are consistent. Third, the results of the simulated-path line of the gas particles in the furnace are also discussed and we find that the results are consistent compared with other research simulation results. Finally, the results of the chemical reaction of nitrogen and carbon oxide are analyzed. The principles of reduced nitrogen oxide pollution are also discussed and the results of the developed model such as nitrogen oxide, carbon dioxide, and carbon oxide mass fraction in the combustion gas are close to the corresponding results of research literature.

9.4.1 The Predicted Temperature Field Analysis

Figure 9.9 shows the temperature distribution in the section which is set longitudinally and diagonally. The temperature is highest in the zone above the area in which the burners are installed. As shown in the figure, the average temperature in this area almost reaches 1750 K, which is close to the corresponding boiler-combustion parameter value in the coal-fired boiler-combustion model developed using CFD technology and based on real data from a power plant [18]. Figure 9.10 shows the temperature distribution in the horizontal and longitudinal section. The trends of the temperature change

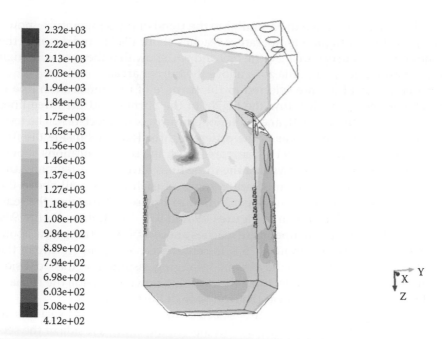

FIGURE 9.9
Contours of temperature field in diagonal-longitudinal section.

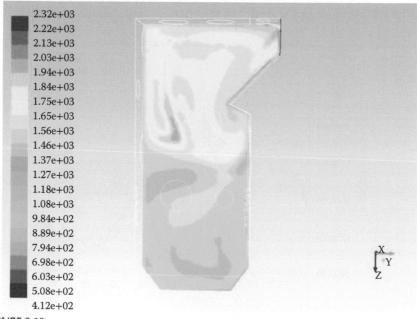

FIGURE 9.10
Contours of temperature field in the horizontal and longitudinal section.

in this section are shown to be similar to the trend of the temperature shown in Figure 9.9. The highest temperature reaches about 1940 K and drops in the zone above the burner area. Figures 9.9 and 9.10 show that the area with high temperature is located in the zone above the burner areas.

Three vertical lines are selected in the furnace of the boiler. Each line is parallel to the z-axis denoting the height of the burner, and the three lines are located in the front, left, and center of the furnace, respectively. The temperature distributions of gas at the different three locations inside the furnace are shown in Figure 9.11. Compared with the burner locations shown in Table 9.5, Figure 9.11 shows that the three lines have similar temperature-change trends. The high-temperature zones of each line are all above the burner areas. The temperature distribution in Figure 9.11 matches the temperature distribution shown in Figures 9.9 and 9.10. In addition, Figure 9.11 shows that the highest temperature reaches about 1950 K in front area, about 1875 K in left area, and about 1725 K in center area. The temperature in the front area is highest due to the assumed slagging assumed on the front side of the furnace, which can decrease the heat flux flowing outside the furnace and raise the temperature near the front side of the furnace.

The three horizontal lines are selected in the latitudinal sections at $z = 20$ m. One of the three lines is from the middle of the left side of the section to the middle of the right side of the section. This is the center line of the section. The other two lines are all parallel to the center line at $y = 4.5$ and 13.5 m, respectively. Figure 9.12 shows the distribution of temperature in the three lines. The trends of temperature change in the three locations are similar. The temperature in the center of the furnace is shown to be higher than

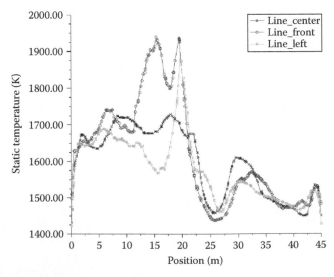

FIGURE 9.11
Temperature distribution along z-axis in the front, left, and center area.

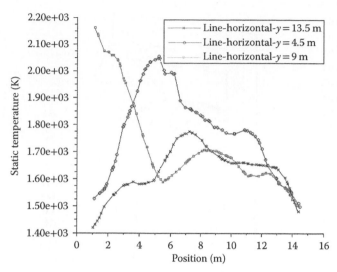

FIGURE 9.12
Temperature distribution parallel to the *x*-axis at *y* = 4.4, 9, and 13.5 m.

temperature in other areas. The highest temperature reaches about 1675 K in the middle of the line *y* = 9 and about 2050 K in the middle of the line *y* = 1775 K and the results are close to the corresponding simulation results 2021 and 1755 K from other researchers [18].

The three different sections are set at *z* = 13, 20, and 25 m, respectively, to show the temperature distribution inside the furnace more completely. Figure 9.13 shows the temperature distribution in the section at *z* = 13 m. The minimum temperature in this section is about 1460 K. The maximum temperature of this section reaches about 2130 K. Figure 9.14 shows the temperature distribution in the section at *z* = 20 m. The minimum temperature in this section is about 1560 K, while the maximum temperature of this section reaches about 2130 K. Both sections are located in the zone above the area of the burners. Figure 9.15 shows the temperature distribution in the section at the level of burner group No. 5. Compared with the sections where *z* = 13 and 20 m, the temperature of the section shown in Figure 9.15 is much lower, and the maximum temperature of this area is just about 1650 K. It is clearly evident that the high-temperature zone is in the zone above the burner-group level. In addition, in the same section, the high-temperature area is always in the center of the section. This can be shown from Figures 9.13 to 9.15. Figure 9.16 shows the temperature distribution inside the furnace. Compared with Figure 9.11, Figure 9.16 shows the temperature distribution in three dimensions using sections in different levels. The temperature distribution shown in Figure 9.16 indicates that the high-temperature area in the furnace is located above the burner area, and the results are the same as the result of the simulation model developed using the same boundary conditions [18].

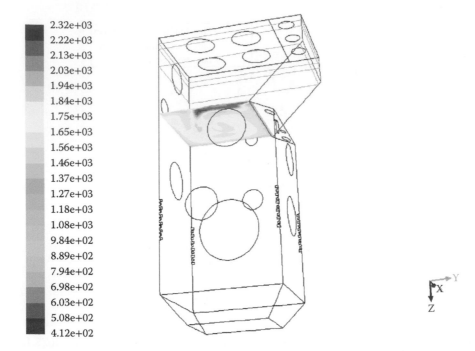

2.32e+03
2.22e+03
2.13e+03
2.03e+03
1.94e+03
1.84e+03
1.75e+03
1.65e+03
1.56e+03
1.46e+03
1.37e+03
1.27e+03
1.18e+03
1.08e+03
9.84e+02
8.89e+02
7.94e+02
6.98e+02
6.03e+02
5.08e+02
4.12e+02

FIGURE 9.13
Contours of temperature at $z = 13$ m.

9.4.2 The Predicted Incident Radiation Analysis

Figure 9.17 shows the incident radiation distribution in the three different lines that are parallel to the z-axis. The trend of the incident-radiation change in the three lines is similar. It is clearly evident that the high-incident-radiation zone is located above the burner area. The maximum incident radiation of the three lines reaches about 3.2×10^5 W/m^2. Figure 9.18 shows the incident radiation distribution in different lines parallel to x-axis. The lines are in different heights with $z = 5, 7, 10, 25$, and 30 m, respectively. It is clearly evident that the high-incident-radiation zone is above the burner area. In addition, at the same height, the incident radiation of the central area is higher than the incident radiation of boundary areas. The maximum incident radiation at the middle line with $z = 7$ m approaches about 3.2×10^5 W/m^2.

9.4.3 The Predicted Gas Particle Trajectory Analysis

Figure 9.19 shows the vectors of velocity magnitude in the horizontal latitudinal section with $z = 13$ m and at the No. 5 burner-group level with $z = 29.315$ m. Tables 9.4 and 9.5 present that the burners in this group blow secondary air with velocity = 54.11 m/s and temperature = 331°C. The turbulence of the gas property inside the furnace is clearly shown. The flue gas at $z = 13$ m swirls

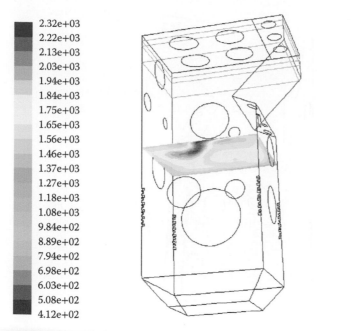

FIGURE 9.14

Contour of temperature distribution in the $z = 20$ m section.

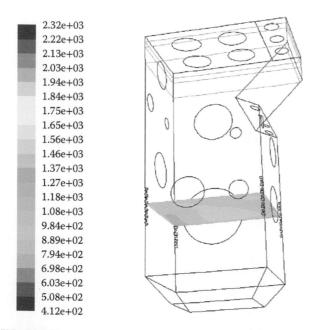

FIGURE 9.15

Contour of temperature distribution in the section at the level of burner group No. 5.

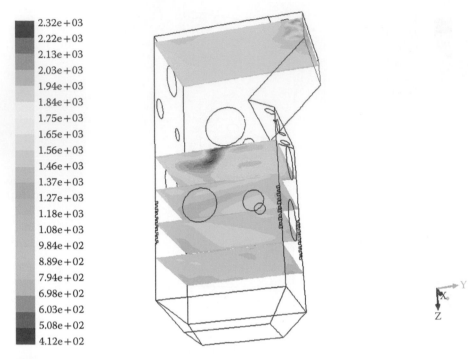

FIGURE 9.16
Contours of temperature field in different latitudinal sections.

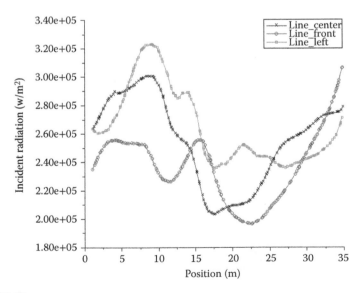

FIGURE 9.17
Incident-radiation distribution along z-axis.

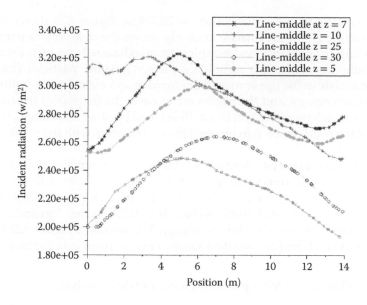

FIGURE 9.18
Incident-radiation distribution along *x*-axis.

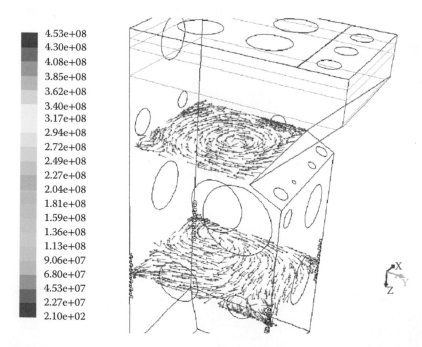

FIGURE 9.19
Vectors of velocity magnitude at *z* = 13 m and level of burner group No. 5.

more drastically than the flue gas at $z = 29.315$ m. Figure 9.20 shows the path line of the flue gas in the furnace. It clearly shows the trajectory of particles of gas entering from burner and escaping from exhausted gas exit. Figure 9.20 also clearly shows the gas turbulence property. Many particles from each burner circulate in the upper area of the furnace and form a vortex with different turbulent energy, and the results are close to the outputs of the coal-fired power plant boiler-combustion model developed by other researchers [112]. It is clear that the results shown in Figure 9.19 are consistent with the results shown in Figure 9.20.

Figure 9.21 shows the gas-density distribution in different lines which are parallel to the z-axis. The trend of the gas-density change of the three lines is similar. The high-density zone is located in the burner area with $z > 27.026$ m and $z < 32.766$ m. The maximum density reaches about 0.235 kg/m³. Figure 9.22 shows the distribution of turbulent kinetic energy. The results of Figure 9.22 indicate that the zone with highly turbulent kinetic energy drops is in the burner area.

9.4.4 The Predicted Nitrogen and Carbon Oxide Analysis

Figure 9.23 shows the mass fraction of the pollutant NO_x and it can be found that the mass of NO_x in the area of the top furnace is very high with a mass fraction that almost reaches 4.00×10^{-6}. Comparing Figure 9.23 with Figures 9.11 and 9.17, it can be found that the mass fraction is higher in

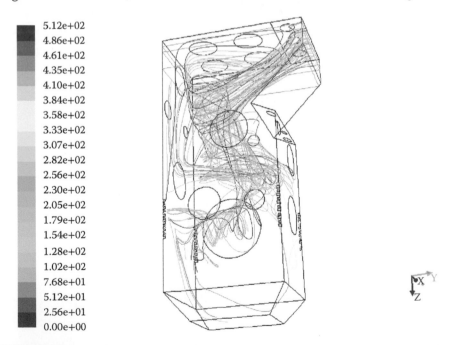

FIGURE 9.20
Path line of flue-gas particles in the furnace.

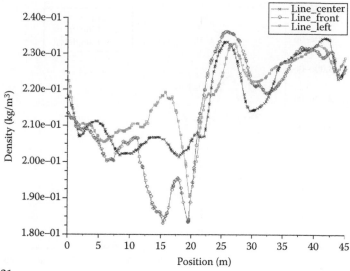

FIGURE 9.21
Gas-density distribution along z-axis in the furnace.

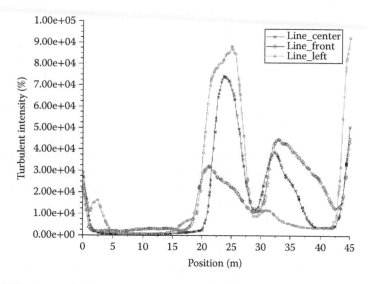

FIGURE 9.22
Turbulence kinetic energy at center, front, back, and right along z-axis.

the area with high temperature and incident radiation. So adjusting input parameters and keeping an appropriate flue-gas temperature can decrease the NO_x emission. This is the mechanism of a low NO_x burner, which has been applied in power plants [130]. In addition, technology is also used to limit NO_x pollution in the power plants without a low NO_x burner by optimizing the air system [130–132].

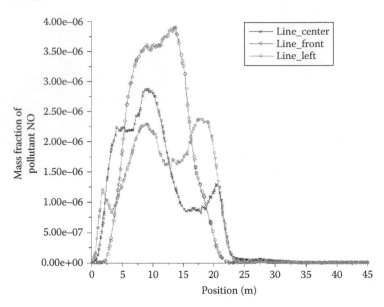

FIGURE 9.23
Contours of mass fraction of NO in front, left, and center of the furnace.

Figure 9.24 shows the distribution of NO along five lines parallel to the x-axis at different height of the furnace. The lines are selected from the upper area and burner area of the furnace. It is clearly evident that the mass fraction of NO is higher in the upper area than the mass fraction of NO in the boundary areas. The maximum mass fraction of NO reaches about 3.4×10^{-6}, and the result is reasonably good [19,20].

Figure 9.25 shows the distribution of NO mass fraction in a horizontal longitudinal section. The distribution perfectly matches the corresponding curves shown in Figure 9.23. Figures 9.23 and 9.25 show that more NO is produced in the zone with $z > 7.5$ m and $z < 17.5$ m. The zone is located above the burner areas with high temperature and incident radiation.

Figure 9.26 shows the distribution of the mole fraction of CO in the furnace. The area with a lower mole fraction of CO is the top of the furnace because carbon is almost completely burnt in this area. The area with a higher mole fraction of CO is close to the burner area where coal starts to burn and a lot of CO is produced. With the gas flowing to the zone above the burner area, the chemical reactions continue and CO converts to CO_2 with significant heat produced [19,20,133]. So, controlling the combustion process and keeping the chemical reaction in the optimal position of the furnace are the main aspects of this research. A chemical reaction kept at the optimal position can maintain the combustion process at a higher heat-transfer efficiency with lower nitrogen and carbon emissions.

Figure 9.27 shows the distribution of CO_2 in the furnace. The area with a high mole fraction of CO_2 is the upper area of the furnace. The mole fraction of CO_2

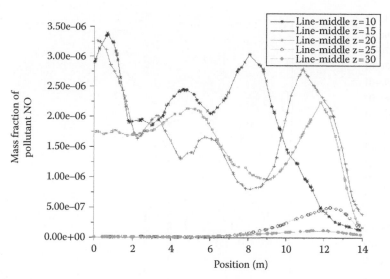

FIGURE 9.24
Distribution of NO in the different heights of the furnace.

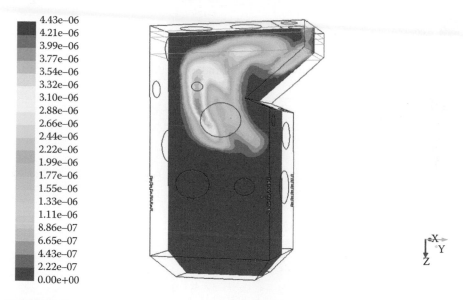

FIGURE 9.25
Contour of NO mass fraction in a horizontal-longitudinal section.

in this area reaches about 0.0947 and the result perfectly matches the results from experiment and the predicted coal combustion in a utility furnace [20].

Figure 9.28 shows the mass fraction of CO_2 along the z-axis. It is found that the mass fraction of CO_2 rises drastically at the position $z = 13$ m, which is the throat of furnace as shown in Figure 9.13. This means that all combustible CO

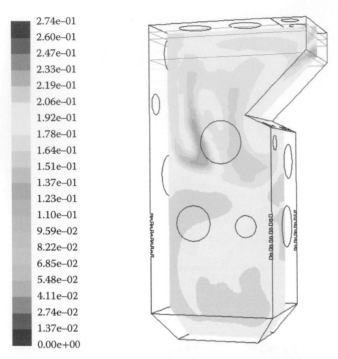

FIGURE 9.26
Contour of mole fraction of CO in horizontal-longitudinal section.

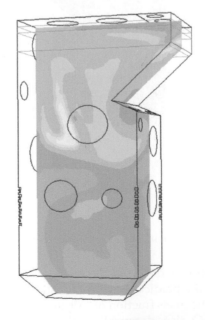

FIGURE 9.27
Contour of mole fraction of CO_2 in horizontal-longitudinal section.

is completely burnt before the gas flows through the throat of the furnace. This is the reason why the temperature and incident radiation are higher above the burner area than the temperature and incident radiation in the other areas.

Figure 9.29 corresponds to Figure 9.26, in which the contour of the mole fraction of CO is shown. Both of them show that the fraction of CO is high

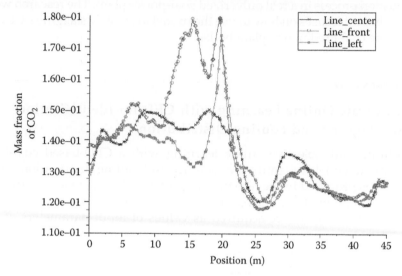

FIGURE 9.28
Mass fraction of CO_2 in the front, center, and right of the furnace along the z-axis.

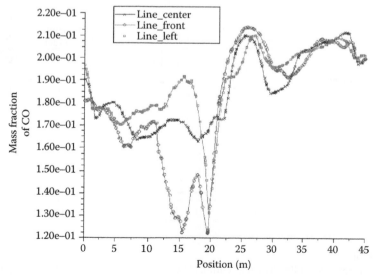

FIGURE 9.29
Mass fraction of CO in the front, center, and right of the furnace along the z-axis.

in the burner area. The carbon rapidly burns starting from the position $z = 25$ m. Compared with the results shown in Figure 9.11, it is clearly evident that the drastically raised temperature leads to much CO being burnt staring from $z = 25$ m.

Therefore, all results of the CFD-based coal-fired power plant boiler-combustion model developed in this research can approximately match the combustion process in a real pulverized coal-power plant. The research work in this chapter focuses on how to use the model to identify slagging distribution in the coal-fired power plant boiler.

9.5 Integrate Online Learning with CFD for Identification of Slagging and Fouling Distribution

Combining multiobjective online learning with a CFD-based coal-fired power plant model to identify the slagging and fouling distribution inside a furnace is proposed in the research. Figure 9.30 shows the logic structure of the proposed method. The real-time process data, such as air and velocity of pulverized coal, are acquired as values of input parameters of the

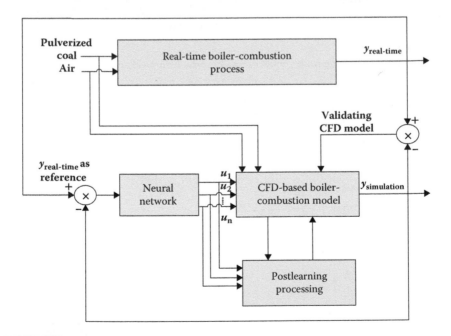

FIGURE 9.30
Logic for integrating online learning with CFD to identify power plant boiler slagging and fouling distribution.

CFD-based model. The CFD model is continuously validated by the error between the condition monitoring data and the corresponding simulation data.

Traditionally, the CFD model is used to simulate the combustion process and keeps the simulation data close to the corresponding real data, so the data that are too difficult to measure from the real power plant process can be easily obtained from the CFD model. This research combines a multiobjective online-learning method with CFD to identify, control, and optimize the real-time power plant-combustion process. Figure 9.30 shows the NN module processes the error between the CFD model output y simulation and real-time output $y_{\text{real-time}}$ and provides output $u_1 - u_n$ as the input to the CFD model. The optimal inputs $u_1 - u_n$ can be found and make the error between $y_{\text{simulation}}$ and $y_{\text{real-time}}$ minimal in the CFD model.

The postlearning processing module is used to decide the specific u_1-u_n by other parameters of the CFD model, because the NN module may provide a number of data pairs of u_1-u_n in which only a part of the pairs can match the CFD model. In the research, the temperature value of the position, where the slagging or fouling is identified, can be used in postlearning processing to validate the u_1-u_n. Figure 9.5 shows a similar situation in which nonequations $x < 69$ and $y < 74$ are used to limited input1 and input2 in a specific part.

9.5.1 Identifying Slagging and Fouling Distribution

A number of circles are set in the CFD-based coal-fired boiler-combustion process model. The circles are set in the position in which the probability of slagging or fouling is high based on the maintenance and operation experience of the power plant. In addition, the diameter of each circle is also a parameter. The circles can work as digital probes. Figure 9.31 shows the digital probes that are set in the position with a high tendency of slagging and fouling.

The software has been developed to realize the proposed research by combining CORBA C++ with ANSYS Fluent 14.5. This is a distributed computing system and can be applied in an environment supported by distributed computing technology such as an Internet-supported environment. Figure 9.32 shows how to apply the system to identify slagging and fouling in a coal-fired power plant. The CFD-based model runs on the server with high performance which is running ANSYS Fluent 14.5. The multiobjective online-learning model can run on another computer with high performance. A CFD-based model is computing intensive because it is supported by the finite element method (FEM). The condition monitoring data can be acquired by the system through the Internet. Figure 9.32 shows how to build the system supported by the Internet, in which the system can provide the slagging and fouling identification service to a number of power plants distributed

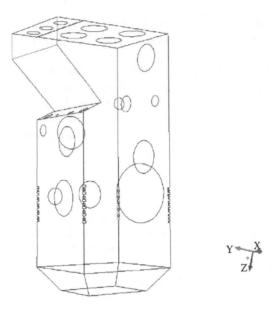

FIGURE 9.31
Digital probes set on the surface of a heat transfer inside furnace for slagging and fouling identification.

in different places simultaneously. Certainly, the system can be applied in a power plant supported by a local area network, and the software structure is the same as the system supported by the Internet.

9.5.2 Analysis of the Results of the Proposed Methodology

A situation with slagging in the front of the furnace is assumed, and the slagging distribution in Figure 9.33 shows that slag is built up in three areas. The smallest one is in the right top of the front side in the furnace, and the heat flux in this area is only about 5000 W. The middle one is located in the left top of the front side in the furnace, and the heat flux in the area reaches about 15,000 W. The biggest one is located in the middle of the front side in the furnace, and the heat flux in the area is about 8000 W.

A server supporting the ANSYS Fluent 14.5 boiler-combustion CFD model and the NN method run on the same personal computer in the research. The CFD-based model is computing intensive, so the amount the software runs is decided by the computer where the combustion model is running. The results of the experiment in this research show that the software can run locally on an industry microcontrol computer to identify the slagging and fouling distribution of a boiler in a coal-fired power plant. However, the Internet-supported distributed computing environment not only can make the identification much faster and accurate, but also can

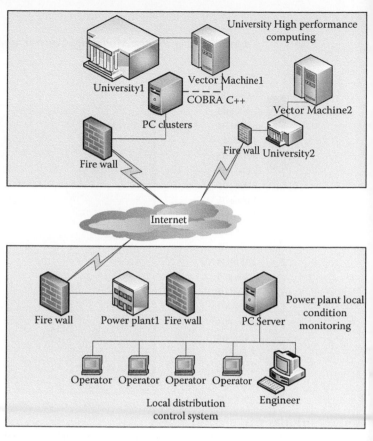

FIGURE 9.32
Distributed computing technology CORBA C++ supported power plant boiler monito optimization.

share the identification service among coal-fired power plants distr in different places.

Figure 9.34 shows the identified slagging distribution using the so of the proposed method. It shows that the system can identify the sl distribution inside the furnace with some errors which can be acc Table 9.7 shows the error of the identification.

Figure 9.35 shows the distribution of surface heat–transfer coeffici real data and the identified data. The trend of the surface heat–transfe ficient change along the z-axis for the real data and the identified similar. The average error reaches about $4\,\mathrm{W/m^2 \cdot k}$. Figure 9.36 sho distribution of surface-incident radiation for the real data and the ide data along the z-axis. The two curves are very approximate in the r > 7.5 m. The errors mainly occur in the range $z < 7.5$ m, and the ma error reaches about 5000 $\mathrm{W/m^2}$. It is clearly evident that the trend

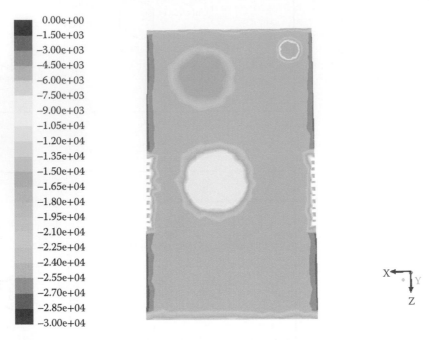

FIGURE 9.33
Slagging distribution on surface of the front side in the furnace.

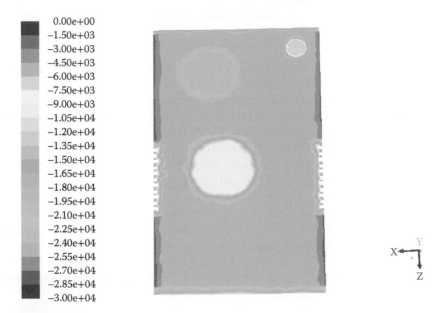

FIGURE 9.34
Identified slagging distribution on surface of the front side in the furnace.

TABLE 9.7

Comparison of Parameters of the Identified and Assumed Slagging

Slagging Area	The Real Heat Flux (W)	The Identified Heat Flux (W)	Heat-Flux Error (W)
Small	5000	6000	1000
Middle	15,000	19,500	4500
Big	8000	7500	−500

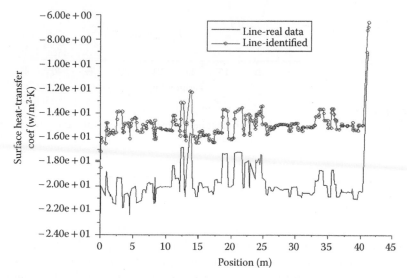

FIGURE 9.35
Surface heat–transfer coefficient of the slagging distribution identified model and the correspondence to real data on the surface of the front side in the furnace.

two curves is similar in Figures 9.35 and 9.36. This means that identification results are close to the real data. Table 9.7 shows that the total heat flux identified is about 5500 W more than the real total heat flux, so the surface heat-flux coefficient and surface incident radiation for the identified are a little higher than the corresponding real data. The results are encouraging although the slagging situation does not exactly match the slagging situation shown in Figure 4.33.

The parameters of the software should be tuned to make the model closer to real practice. In addition, a microcomputer with high computing performance can improve the accuracy of the software because the model can iterate many more times and get more satisfying approximation.

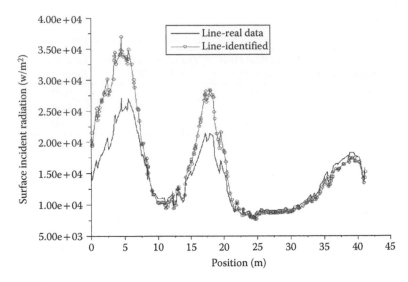

FIGURE 9.36
Surface incident radiation of the identified slagging-distribution model and the corresponding real data on the surface of the front side in the furnace.

9.6 Conclusion

Slagging and fouling is one of the most severe combustion problems occurring in a coal-fired power plant with coal quality frequently changing and significant errors that vary from the design-rated value. In addition, the frequently varying conditions inside the furnace can also lead to slagging in a coal-fired power plant. The efficiency can be seriously deteriorated by slagging and fouling built up on the heat-transfer surface. This research has proposed a new method to identify slagging and fouling distribution and quantify slagging and fouling by integrating multiobjective online learning with CFD technology, and the practical software is developed to realize the proposed method. In addition, a practical computing environment supported by the Internet and higher performance computing technologies is recommended to apply the research to industry. With the advancement of the Internet and computing technologies, the fossil fuel–combustion process can be simulated accurately using a CFD-based model. Furthermore, the conventional combustion process control, identification, and optimization can be combined with CFD to provide a new effective way to practice. Undoubtedly, with this new method the fossil fuel power-generation industry can more efficiently produce electricity to meet regulated carbon emissions. Further methods of how to combine multiobjective online learning with CFD technology to improve thermal power plant boiler efficiency are discussed in Chapter 10.

10

Integrating Multiobjective Optimization with Computational Fluid Dynamics to Optimize the Boiler-Combustion Process

10.1 Introduction

Coal-fired power plant boiler combustion is a highly complex process, and improving the combustion process requires multiobjective optimization. A combustion process with two objectives is shown in Figure 5.1, where Q_1 is the total heat absorbed by the tubes of heat-transfer equipment inside a boiler such as a water wall, superheater, reheater, and economizer, and maximum Q_1 is expected. Normally, if the temperature of the flue gas is higher than $T_{melting}$, which is the ash-melting temperature shown in Figure 5.1, the ash starts to melt and slagging increases. Therefore, an efficient boiler-combustion process should have maximum Q_1 with flue-gas temperature in the areas close to the sides of the furnace under $T_{melting}$.

Artificial intelligence (AI) technologies, such as the neural network–based methods and multiobjective optimization, have been applied in industry to improve the efficiency of control systems [8,9,73,82–85,107]. For example, neural-network-driven computer systems are used to optimize soot blowing in a coal plant boiler, reduce NO_x emissions, improve heat rate and unit efficiency, and reduce particulate matter emissions in coal-fired power plants in the United States [5]. Nondominated sorting in genetic algorithm (NSGA II) is one of the AI-based multiobjective optimizations and is widely used to successfully optimize industry processes [86–93]. In addition, computational fluid dynamics (CFD) simulation technology is widely applied in the power-generation industry to analyze combustion process [94,95], improve boiler design [96–100], and adjust burner-tilt angle in offline fashion after an overhaul or upgrade at a power plant [101].

In fact, with the advancement of computer technology and mathematical methodology, integrating AI with CFD technologies can solve combustion-related problems. Based on this, the research proposes new methods to improve combustion-process efficiency and decrease carbon emissions for fossil fuel power-generation industry by integrating multiobjective optimization with CFD technology to improve boiler-combustion efficiency and decrease or even prevent serious slagging inside the furnace of a coal-fired power plant boiler.

10.2 Principle Mechanism of Combustion Process and Slagging inside a Coal-Fired Power Plant Boiler

10.2.1 The Heat-Transfer Process inside a Boiler

Three modes of heat transfer occur inside a boiler: radiation, convection, and conduction. Radiant heat transfer is prevalent in the radiation section or core of the furnace and the transfer of radiant energy to the boiler tubes is dependent on the luminosity of the flame and the amount of heat-absorbing surface of the boiler [102]. The governing equation of the rate of radiant heat absorbed by the water wall is given by [19]

$$q_{rad} = \varepsilon A F \sigma \left[T_1^4 - T_2^4 \right] \tag{10.1}$$

where ε is the emissivity of the flame, A is the area of a cross-section of radiant heat surface m^2, F is the view factor, σ is the Stefan–Boltzmann constant 5.67 × 10^{-8} W/m^2·K^4, T_1 is the absolute temperature of flame K, and T_2 is the absolute temperature of radiant heat absorbing surface K. Heat transfer in the fluid path area takes place entirely by convection. The rate of the heat transfer from the flue gas to the heat absorbing surface laid out in this zone is given by [19,21]

$$q_{conv} = h_{conv} A \Delta\theta \tag{10.2}$$

where h_{conv} is the coefficient heat transfer W/m^2·K, A is the area of heat absorbing surface m^2, and $\Delta\theta$ is the temperature difference between the fluid gas and surface of the metal tube of heat-transfer equipment such as the superheater, reheater, economizer, and air reheater. Heat transfer by the mode of conduction takes place through the wall thickness of tubes as well as across the slag deposited on the surface of the water wall. The rate of conductive heat transfer through a water wall is given by [19,20,81]

$$q_{cond} = kA \frac{\Delta\theta}{\Delta x} \tag{10.3}$$

For a composite wall, we have

$$q_{cond} = k_1 A \left[\frac{\Delta\theta}{\Delta x} \right]_1 + k_2 A \left[\frac{\Delta\theta}{\Delta x} \right]_2 \tag{10.4}$$

where k_1 and k_2 are the thermal conductivities of slag deposition and metal tubes of water walls (W/m·K), and $\left[\dfrac{\Delta\theta}{\Delta x}\right]_1$ and $\left[\dfrac{\Delta\theta}{\Delta x}\right]_2$ are the temperature gradients across the slag and the tubes of water walls (K/m).

All three heat transfer processes take place simultaneously in the boiler-combustion process.

10.2.2 The Predicted Temperature Field Analysis

As the pulverized coal is blown into the furnace and heated, the volatile matter in the coal is liberated. As the bonds between the coal molecules are broken, the coal decomposes many gases that are contained in the gaseous volatiles including CO_2, H_2O, N_2 and small proportions of CO, H_2, and many different hydrocarbons (C_xH_y). These are mixed with the surrounding air and rapidly burnt with the prevailing temperature above the ignition temperature of the volatile mixture [102]. The important chemical reactions that take place inside coal-fired boiler combustion are shown [102,19,21]:

$$H_2 + \frac{1}{2}O_2 = H_2O \text{ (vapor)} \tag{10.5}$$

$$C + \frac{1}{2}O^2 = CO \text{ (gas)} \tag{10.6}$$

$$CO + \frac{1}{2}O^2 = CO_2 \text{ (gas)} \tag{10.7}$$

$$CO_2 + C = 2CO \text{ (gas)} \tag{10.8}$$

$$C + H_2O = CO + H_2 \text{ (gas)} \tag{10.9}$$

$$CH_4 + O_2 = 2CO_2(gas) + H_2 \text{ (vapor)} \tag{10.10}$$

$$C_2H_2 + 2.5O_2 = 2CO_2(gas) + H_2O \text{ (vapor)} \tag{10.11}$$

Equation 10.5 shows the hydrogen contained in the coal reacts with oxygen in the air. Equations 10.6 and 10.7 show the carbon contained in the coal reacts with the oxygen of air in the furnace. This is the main source of heat liberated from pulverized coal burning. Equation 10.8 shows that carbon dioxide can react with carbon and release carbon monoxide. Equation 10.9 shows that vapor can react with carbon and hydrogen and carbon monoxide are produced. Equations 10.10 and 10.11 show that the hydrocarbon reacts with oxygen and carbon oxides are released.

10.2.3 The Mechanisms of Slagging in the Coal-Fired Boiler

Slagging in the radiant section of the coal boiler with a high temperature is usually associated with some degree of melting of the ash [102]. In coal-fired power plant boilers, slagging can occur on the furnace water walls and the first few rows of superheater tubes. The aerodynamics of the flue gas in the combustion process can convey ash particles to the vicinity of the heat transfer, and the ash particles can pass to the boundary area by inertia. Figure 5.1 shows the boundary area L_{right} which is close to the right side of the furnace. The ash particles can adhere to the surface of water wall tubes if either the particles or the surface is "sticky" enough to overcome the kinetic energy of the incoming particles and prevent it from rebounding from the heat-transfer surface [102]. Therefore, maintaining an appropriate temperature in the boundary of the furnace and keeping the incoming particles from melting can decrease slagging. Based on these mechanisms of heat transfer, chemical reactions, and slagging, the research proposes a novel way to improve coal-fired power plant boiler efficiency and decrease slagging.

10.3 Modeling of Coal-Fired Power Plant Boiler-Combustion Process

The CFD-based model of the coal-fired power plant boiler-combustion process shown in Figure 5.1 is created in this research. Equation 5.1 gives an expression of the heat balance in the coal boiler-combustion process shown in Figure 5.1. Q_1 is normally in the range of 75%–90% [102]. However, slag, which is accumulated on the heat-transfer surface, can seriously influence Q_1. The thermal efficiency of the coal-fired boiler can be expressed as [102]

$$\eta = \frac{Q_{steam}}{Q_{coal}} \tag{10.12}$$

where Q_{steam} is the useful heat out in steam, and Q_{coal} is the heat in from coal:

$$Q_{coal} = Q_1 + Q_2 + Q_3 + Q_4 + Q_5 + Q_6 \tag{10.13}$$

where Q_1 Q_6 are the same as denoted in Equation 5.1.

The fields of temperature, pressure, velocity, and density of flue gas inside a coal boiler are dynamic and Equations 6.1 through 6.4 cannot be used to predict all the fields of flue gas accurately. Therefore, the finite element method supported with CFD is applied to simulate all the dynamic fields of the flue gas more accurately [20–22,54,81].

A three-dimensional power plant boiler furnace model is developed using ANSYS Fluent 14.5 based on the real data, which is from a 1160 t/h tangential coal-fired power plant [54]. The characteristics of flue-gas property fields, such as temperature and intensity, are analyzed below and the results show that the simulation results of the flue-gas properties are close to the corresponding data from the power-generation industry and simulation results from research [21], [81]. The geometry model is developed based on the data from the power plant [54]. This is a 14.62 m wide, 12.43 m deep, and 48.8 m high furnace with tangential-combustion fashion. The geometry of the furnace is shown in Figure 6.41.

The positions of each set of four burners located in the same horizontal section are shown in Figure 6.42. The center line of the burner, which is installed in different corner, is shown in Figure 6.43. The mesh for the geometry is shown in Figure 6.44. CORBA C++ is used to integrate ANSYS Fluent 14.5 with a multiobjective-optimization model developed using MATLAB. The details of the CFD-based coal-fired boiler-combustion model such as geometry data and combustion-model equations and data can be referenced in Section 4.3.

10.4 NSGA II-Based Multiobjective Optimization Model

A modified NSGA II is a widely applied nondomination-based multiobjective-optimization method. The steps of how to use NSGA II to solve problems regarding multiobjective optimization are presented below. First, the population number is set to define a chromosome group. Each chromosome is composed of many traits that correspond to the decision variables practical problem domain in the research. The values of velocity for two sets of burners and the values of temperature for each set of burners are defined as the decision variables. Table 10.1 shows the mapping between the traits of each chromosome and the decision variables of the problem domain.

TABLE 10.1

Mapping between Chromosome Traits and Decision Variables of the Problem Domain

Decision Variables	Min. Value	Max. Value	Traits of a Chromosome
Velocity of primary air A3 (m/s)	20	65	Trait 1
Velocity of primary air B3 (m/s)	20	65	Trait 2
Velocity of primary air C3 (m/s)	20	65	Trait 3
Velocity of primary air D3 (m/s)	20	65	Trait 4
Velocity of secondary air A3 (m/s)	55	85	Trait 5
Velocity of secondary air B3 (m/s)	55	85	Trait 6
Velocity of secondary air C3 (m/s)	55	85	Trait 7
Velocity of secondary air D3 (m/s)	55	85	Trait 8
Temperature of burners set 6 (K)	400	575	Trait 9
Temperature of burners set 5 (K)	500	675	Trait 10

Second, a nondominated sort is carried out on the chromosome group with the population assigned in the first step. The principles of the sort are based on the values of each objective function and the crowding distance of each chromosome in the group. The value of the objective has higher priority than the crowding distance in the sort. Two objective functions shown in Table 10.2 are set in this research.

Objective function 1 is used to maintain the coal boiler to run at a higher heat-transfer rate. Objective function 2 is used to control the temperature so that it is not enough for the ash particles to be melted in the areas which are close to the sides of furnace, and the ash particles do not become sticky. This can decrease the trend of slagging on the surface of the water wall. The crowding distance is calculated as follows [100]:

$$I(d_{k+1}) = I(d_k) + \frac{I(k+1)_m - I(k-1)_m}{f_m^{max} - f_m^{min}} \tag{10.14}$$

where $I(d_1) = \infty$, $I(d_n) = \infty$, $I(k)_m$ is the value of the mth objective function of the kth individual in group I, and f_m^{max} and f_m^{min} are the maximum and minimum value of the mth objective function, respectively. Third, the group of chromosomes is processed using genetic operators, and the new chromosomes are produced. The binary crossover observed in nature is given as [134]

$$c_{1,k} = \frac{1}{2}[(1 - \beta_k)p_{1,k} + (1 + \beta_k)p_{2,k}] \tag{10.15}$$

$$c_{2,k} = \frac{1}{2}[(1 + \beta_k)p_{1,k} + (1 - \beta_k)p_{2,k}] \tag{10.16}$$

where $c_{i,k}$ is the ith child with k^{th} component, $p_{i,k}$ is the selected parent, and $\beta_k(\geq 0)$ is a sample from a random number general generated having the density

$$p(\beta) = \frac{1}{2}(\eta_c + 1)\beta^{\eta_c}, if\, 0 \leq \beta \geq 1 \tag{10.17}$$

$$p(\beta) = \frac{1}{2}(\eta_c + 1)\frac{1}{\beta^{\eta_c + 2}}, if\, 0 \leq \beta \geq 1 \tag{10.18}$$

TABLE 10.2

Objective Functions

Item Number	Description of Objective Functions
1	Maintaining the coal boiler to run at a higher heat-transfer rate
2	Keeping the temperature of the areas which are close to the sides of the furnace lower than the ash-melting temperature

where η_c is the distribution index for the crossover. The distribution is given [100] as follows:

$$\beta(u) = (2u)^{\frac{1}{\eta+1}}$$ (10.19)

where u is the random number in the range (0, 1). One of the genetic operation polynomials is given [134] as follows:

$$c_k = p_k + (p_k^u - p_k^l)\delta_k$$ (10.20)

where c_k is the child and p_k is the parent with p_k^u the upper bound on the parent component, p_k^l the lower bound, and δ_k a small variation:

$$\delta_k = (2r_k)^{\frac{1}{\eta_m+1}} - 1, \text{ if } r_k < 0.5$$ (10.21)

$$\delta_k = 1 - [2(1 - r_k)]^{\frac{1}{\eta_m+1}}, \text{ if } r_k \geq 0.5$$ (10.22)

where r_k is a uniformly sampled random number in the range (0, 1), and η_m is mutation-distribution index. Finally, the new offspring group is created, and the same processing is carried out on the new generation. This iteration does not stop until the expected values of the objective functions are obtained.

10.5 Integrating the NSGA II Multiobjective-Optimization Method with CFD to Optimize the Coal-Fired Power Plant Boiler-Combustion Process

The coal-fired power plant combustion process is too complex to get analysis equations from which the objective functions regarding characteristics of the fields of flue-gas temperature, pressure, velocity, and density can be obtained. The lack of objective functions can make it difficult to apply multiobjective optimization in boiler-combustion efficiency improvement. This research proposes a novel method to improve the boiler-combustion efficiency by integrating multiobjective optimization with CFD. Figure 10.1 shows the logic of the proposed method.

In Figure 10.1, the multiobjective-optimization module gets the values of objective functions. Then the module provides a vector of decision variables u_1–u_n. The decision variables vector feeds the CFD and new simulation outputs are produced such as dynamic fields of flue-gas temperature, pressure, velocity, density, and heat-transfer rate. The values of these dynamic characteristics are regarded as objective functions and fed back to the

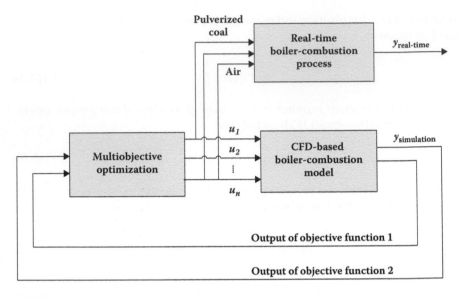

FIGURE 10.1
Logic of integrating multiobjective optimization with CFD to improve coal-fired power plant-boiler efficiency.

multiobjective-optimization module again. Like an optimization method with an analysis objective function, the optimal decision variables can be obtained after the required iterations.

From the principles of heat transfer and the mechanism of slagging in a coal-fired power plant boiler, the appropriate fields of flue-gas temperature, pressure, velocity, and density can not only maintain the boiler at a higher heat-transfer rate but also decrease or even avoid slagging. The velocity of two sets of burners and the temperature of the primary and secondary air are selected as decision variables in this research. The details of the decision variables are shown in Table 10.3. The range of each decision variable is also provided in the table. The two objective functions are provided by the CFD-based coal-boiler-combustion model. Getting a higher value of the total heat-transfer rate of the boiler sides is set as the first objective function. The objective function is formulated as

$$max \ f = \eta \tag{10.23}$$

$$\text{constraints: } v_{min} \leq v \leq v_{max}, \text{ and } T_{min} \leq T \leq T_{max}$$

where η is the total heat flux of sides in furnace, v_{min} and v_{max} are the minimum and maximum values for the velocity of each burner, and T_{min} and T_{max} are the minimum and maximum value of the air temperature for each burner. The second objective function is to keep the temperature in the area which is

TABLE 10.3

Objective Functions under Unoptimized Decision Variables

Decision Variables/Objective Functions	Minimum Value	Maximum Value	Unoptimized Decision Variables
Velocity of primary air A3 (m/s)	20	65	42.5
Velocity of primary air B3 (m/s)	20	65	42.5
Velocity of primary air C3 (m/s)	20	65	42.5
Velocity of primary air D3 (m/s)	20	65	42.5
Velocity of secondary air A3 (m/s)	55	85	70
Velocity of secondary air B3 (m/s)	55	85	70
Velocity of secondary air C3 (m/s)	55	85	70
Velocity of secondary air D3 (m/s)	55	85	70
Temperature of burners set 6 (K)	400	575	487.5
Temperature of burners set 5 (K)	500	675	587.5
Average heat-transfer rate	Total heat-transfer rate of the boiler sides = 82,882,749 W/(m²·K)		
Keeping the temperature of the areas which are close to the sides of the furnace lower more than the ash-melting temperature	The difference between the gas temperature of the cared area and the ash-melting temperature set in the research is 861.647 K		

close to the sides of furnace lower than the melting temperature of the ash. The objective function is formulated as

$$|T - T_{melt}| < \varepsilon \tag{10.24}$$

where T is temperature, T_{melt} is the ash-melting temperature, and ε is the desired difference between the temperature and the ash-melting temperature. T_{melt} is set as 1500 K in this research [99]. Figure 10.2 shows that the horizontal lines are set in the areas which are close to the sides of the furnace. The temperature along the lines is controlled and maintained not to be more than the melting temperature at which ash can melt and become sticky. Figure 10.3 shows the vertical lines set on the sides of the furnace, and the temperature along the lines can be analyzed.

Two scenarios are developed and analyzed in the research. In the first scenario, the decision variables shown in Table 10.3 are not optimized, and the values of velocity for each burner are set to the average value of each burner's allowed input range. The values of the temperature of the primary and secondary air are also set to the average value of their allowed input range. The results of the objective functions in the first scenario are shown in Table 10.3. The total heat-transfer rate of the boiler water wall without combustion optimization is 82,882,749 W/(m²·K). The difference between the flue maximum flue-gas temperature and the ash-melting temperature in the areas which are close to the sides of furnace is 861.647 K. Figure 10.4 shows the temperature distribution in the back sides of the furnace. The temperature in the back side of the furnace is in the range of 1480–1880 K.

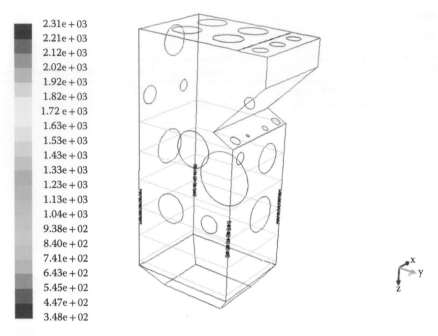

FIGURE 10.2
Horizontal lines set in the areas which are close to the side of the furnace.

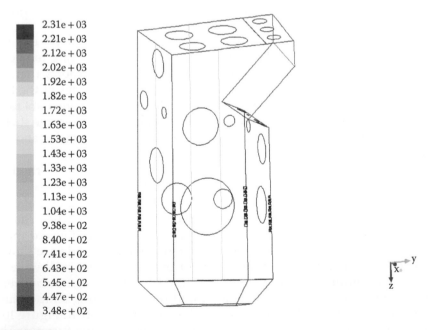

FIGURE 10.3
Vertical lines set on the sides of the furnace.

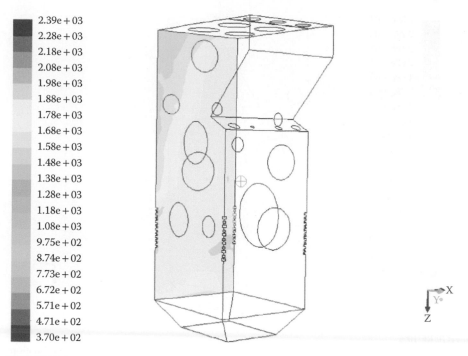

2.39e+03	
2.28e+03	
2.18e+03	
2.08e+03	
1.98e+03	
1.88e+03	
1.78e+03	
1.68e+03	
1.58e+03	
1.48e+03	
1.38e+03	
1.28e+03	
1.18e+03	
1.08e+03	
9.75e+02	
8.74e+02	
7.73e+02	
6.72e+02	
5.71e+02	
4.71e+02	
3.70e+02	

FIGURE 10.4
Temperature distribution on the back side of the furnace without boiler-combustion optimization.

Figure 10.5 shows the temperature distribution on the right side of the furnace in the first scenario. The temperature in the area is in the range of 1580–1980 K. Figure 10.6 shows the temperature distribution along the vertical lines set in the middle of each side of the furnace in the first scenario. The temperature approaches about 1800 K in most areas of the lines, and the value of temperature is consistent with the value of temperature in Figures 10.4 and 10.5. Figure 10.7 shows the temperature distribution in the horizontal lines in the back side of the furnace at $z = 15, 20, 25, 35,$ and 40 m in the first scenario. The temperature in this area is more than 1780 K, and the maximum temperature reaches about 1890 K.

Figure 10.8 shows the temperature distribution along the horizontal lines on the left side of the furnace at $z = 15, 20, 25, 35,$ and 40 m in the first scenario. The temperature distribution in each denoted level of the z-axis is more than the ash-melting temperature 1500 K defined in the research. Figure 10.9 shows the temperature distribution along the horizontal lines on the front side at $z = 15, 20, 25, 35,$ and 40 m in the first scenario. The average temperature in the side is about 1800 K beyond the ash-melting temperature of 1500 K set in the research. Figure 10.10 shows the temperature distribution along the horizontal lines on the right side of the furnace. The minimum temperature

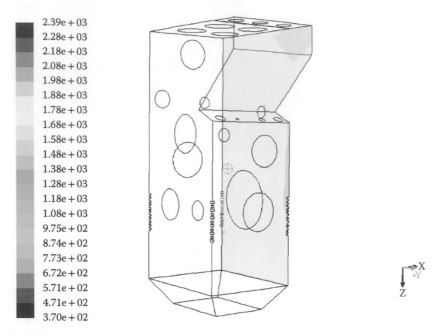

FIGURE 10.5
Temperature distribution on the right side of the furnace without boiler-combustion optimization.

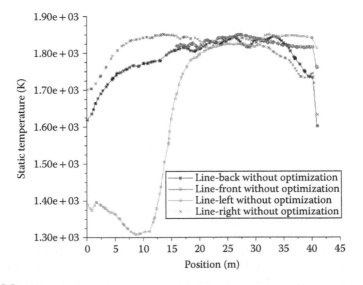

FIGURE 10.6
Temperature distribution along the vertical lines set on the sides of the furnace without boiler-combustion optimization.

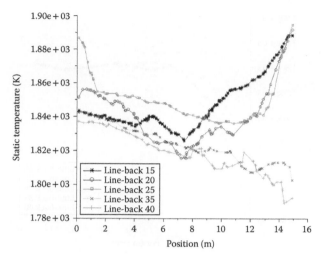

FIGURE 10.7
Temperature distribution along horizontal lines set on the back side of the furnace at $z = 15, 20, 25, 35,$ and 40 m without boiler-combustion optimization.

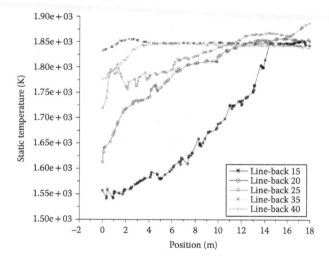

FIGURE 10.8
Temperature distribution along horizontal lines set on the left side of the furnace at $z = 15, 20, 25, 35,$ and 40 m without boiler-combustion optimization.

of the side is about 1750 K and the maximum temperature of the side reaches about 1890 K. The coal-fired boiler combustion in the first scenario is not optimized and it is observed that the temperature in the areas close to the sides of the furnace is far more than the ash-melting temperature of 1500 K.

In the second scenario, the coal-boiler combustion is optimized and Table 10.4 shows the optimized decision variables and the results of the

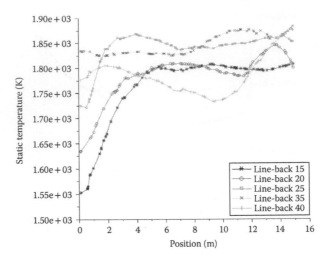

FIGURE 10.9
Temperature distribution along horizontal lines set on the front side of the furnace at $z = 15$, 20, 25, 35, and 40 m without boiler-combustion optimization.

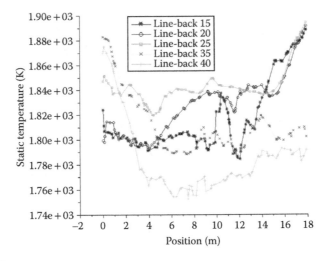

FIGURE 10.10
Temperature distribution along horizontal lines set on the right side of the furnace at $z = 15$, 20, 25, 35, and 40 m without boiler-combustion optimization.

objective functions. The total heat-transfer rate of the sides of furnace is 83,254,837 W/(m²·K). The difference between the flue-gas temperature and the ash-melting temperature is 46.72 K. Compared with the results in the first scenario, the optimized decision variables applied in the second scenario can not only maintain the temperature in the areas close to sides of furnace within the ash-melting temperature limit but also obtain a total heat-transfer

TABLE 10.4

Objective Functions under Optimized Decision Variables

Decision Variables/Objective Functions	Minimum Value	Maximum Value	Optimized Decision Variables
Velocity of primary air A3 (m/s)	20	65	27.4805
Velocity of primary air B3 (m/s)	20	65	49.837
Velocity of primary air C3 (m/s)	20	65	40.2855
Velocity of primary air D3 (m/s)	20	65	35.8455
Velocity of secondary air A3 (m/s)	55	85	56.7112
Velocity of secondary air B3 (m/s)	55	85	73.2305
Velocity of secondary air C3 (m/s)	55	85	78.4996
Velocity of secondary air D3 (m/s)	55	85	79.0782
Temperature of burners set 6 (K)	400	575	490.979
Temperature of burners set 5 (K)	500	675	552.841
Maintaining the coal boiler to run at higher heat-transfer rate	Total heat-transfer rate of the boiler sides = 83 254 837 W/(m²·K)		
Keeping the temperature of the areas which are close to sides of furnace lower than the ash-melting temperature	The difference between the gas temperature of the cared area and the ash-melting temperature set in the research is 46.72 K		

rate of the sides which is similar to the one obtained in the first scenario. Maintaining the temperature lower than the ash-melting temperature in the areas which are close to the sides of the furnace can decrease or even prevent slagging on the heat surface of the water wall, and this is significantly beneficial to the coal-fired power plants in which coal quality frequently changes and there is a high trend of slagging.

Figure 10.11 shows the temperature distribution on the back side of the furnace in the second scenario. The maximum temperature of the side is about 1460 K. Figure 10.12 shows the temperature distribution on the right side of the furnace. The temperature on this side is lower than the ash-melting temperature of 1500 K.

Compared with the corresponding temperature distribution in the first scenario in which coal boiler combustion is not optimized, the temperature distribution in the areas close to the sides of the furnace of the boiler in Figures 10.11 and 10.12 is maintained in an appropriate range. Figure 10.13 shows the temperature distribution along the vertical lines on the sides. It is observed that the maximum temperature on the sides is about 1550 K, which is close to the ash-melting temperature 1500 K.

Figure 10.14 shows the temperature distribution on horizontal lines set in the area which is close to the back side of the furnace in the second scenario. The temperature distribution is in the range of 960–1015 K. The temperature falling in this range cannot cause slagging on the water wall of the boiler. Figure 10.15 shows the temperature distribution on the horizontal lines in the area which is close to the left side of the furnace. The maximum temperature along the line with $z = 15$ m reaches about 2200 K, and the maximum

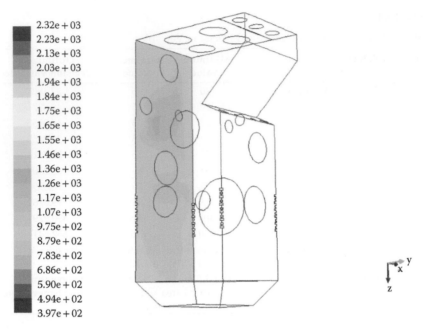

FIGURE 10.11
Temperature distribution on the back side of the furnace with boiler-combustion optimization.

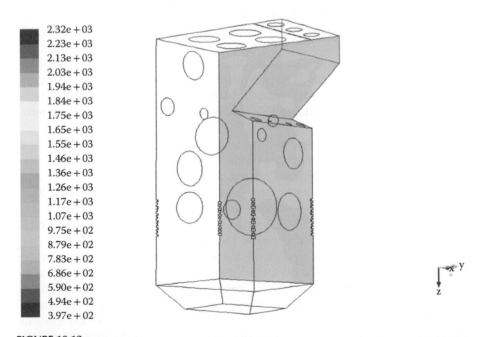

FIGURE 10.12
Temperature distribution on the right side of the furnace with boiler-combustion optimization.

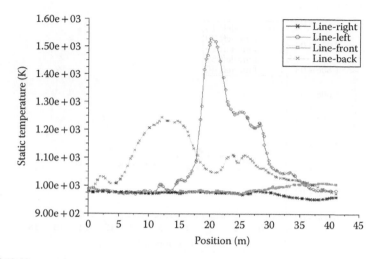

FIGURE 10.13
Temperature distribution along the vertical lines set on the sides of the furnace with boiler-combustion optimization.

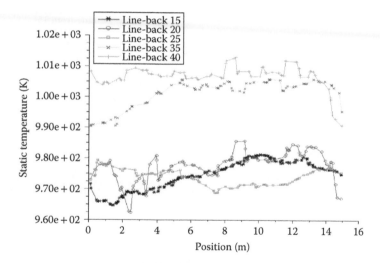

FIGURE 10.14
Temperature distribution along horizontal lines set on the back side of the furnace at $z = 15$, 20, 25, 35, and 40 m under boiler-combustion optimization.

temperature along the line with $z = 20$ m is about 1650 K. Both temperatures in most parts of the two areas are more than the ash-melting temperature. However, the temperature along the other lines on the sides is lower than the ash-melting temperature.

Figure 10.16 shows the temperature distribution along the horizontal lines in the area which is close to the front sides of the furnace in the second

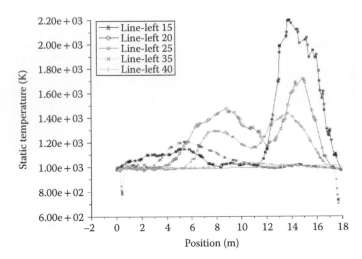

FIGURE 10.15
Temperature distribution along horizontal lines set on the left side of the furnace at $z = 15$, 20, 25, 35, and 40 m under boiler-combustion optimization.

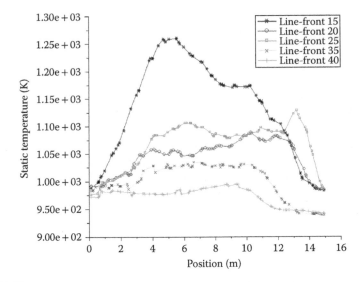

FIGURE 10.16
Temperature distribution along horizontal lines set on the front side of the furnace at $z = 15$, 20, 25, 35, and 40 m under boiler-combustion optimization.

scenario. The temperatures along all lines are lower than 1300 K. Figure 10.17 shows the temperature distribution along the horizontal lines on the right side of the furnace. The maximum temperature along the lines on the side is about 1015 K. Compared with the corresponding temperature distribution in the first scenario, the optimization can maintain the temperature in Figures 10.16 and 10.17 in an appropriate range on which ash is not hot

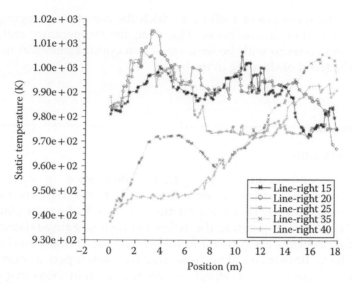

FIGURE 10.17
Temperature distribution along horizontal lines set on the right side of the furnace at $z = 15, 20,$ 25, 35, and 40 m under boiler-combustion optimization.

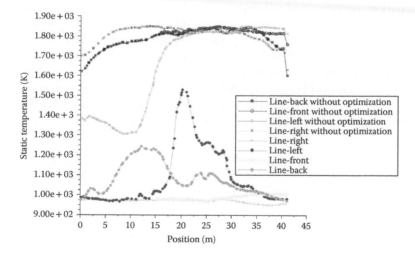

FIGURE 10.18
Comparison of the temperature distribution along lines set on the sides of the furnace with and without boiler-combustion optimization.

enough to melt and stuck on the heat surface of the water wall. Figure 10.18 shows the comparison of the temperature distribution along vertical lines set in the middle of the sides of the furnace both in the first and second scenario. The temperature in the areas which are close to the sides of the furnace of the first scenario is much higher than the temperature in the cared areas of the second scenario. The temperature in the first scenario is far beyond the ash-melting temperature of 1500 K at which the ash starts slagging on the heat-transfer surface in the boiler. However, the temperature maintained in the second scenario with boiler-combustion optimization can massively decrease the trend of slagging in the boiler.

10.6 Conclusion

Optimizing coal-fired power plant boiler combustion is significant to improve power plant efficiency and decrease carbon emissions. However, it is difficult to optimize boiler combustion for coal-fired power plants with high trends of slagging inside the boiler because slagging-related boiler-combustion problems are difficult to solve as AI-based boiler-optimization methods have limitations. This research has developed a new boiler-combustion optimization method of integrating multiobjective optimization with CFD. This method can optimize slagging-related coal-boiler combustion in which conventional methods cannot work successfully because of limitations caused by the lack of instrument data from the boiler-combustion process and slagging. The boiler-combustion process optimized using this method can not only maintain a higher heat transfer of the water wall but also keep the temperature in areas close to the sides of the furnace of the boiler within the ash-melting temperature limit. Controlling the temperature of the areas close to the sides of the furnace can effectively decrease or even avoid slagging and this can massively improve boiler efficiency. Furthermore, the software has been developed using CORBA C++ combined with ANSYS Fluent 14.5 to realize the method.

Part IV

Thermal Power Plant Optimization Solution Supported by High-Performance Computing and Cloud Computing

11

Internet-Supported Coal-Fired Power Plant Boiler Combustion Optimization Platform

11.1 Introduction

A conventional coal-fired power plant computer system is shown in Figure 11.1. The computer system is divided into five levels. Level 1 is the equipment level in which the machines need to be optimized. Level 2 is the controller level in which the intelligent devices are set to control the machines of level 1. Level 3 is the process control level in which each process with a specific function is controlled. Level 4 is the condition monitoring and optimization level in which all the real-time processes located in level 3 are integrated, supervised, and optimized. The proposed identification, control, and optimization methods in the research can be applied in level 4. Level 5 is the management information level including asset management, finance management, and operation management. The computer system shown in Figure 11.1 is a local computer system. The condition monitoring and optimization system cannot be shared by another power plant. In addition, the different power plants cannot share information such as identification or optimization results among the different computer systems. Furthermore, the hardware, such as server machines and other devices of the computer system, cannot be used to support more than one power plant local computer system.

However, with the advancement of Internet-based technologies, such as cloud computing, web services, and CORBA technologies [135–141], a coal-fired power plant boiler combustion optimization platform supported by the Internet can be built to provide identification and optimization services for all power plants distributed in different places. This research proposes machine learning integrated with a computational fluid dynamics (CFD)-based method to identify, control, and optimize coal-fired power plant boiler combustion and all the methods have been developed using CORBA C++ combined with ANSYS Fluent 14.5. The developed furnace slagging identification and boiler combustion optimization software can not only be used in a local computer system but can also be applied in the Internet-supported

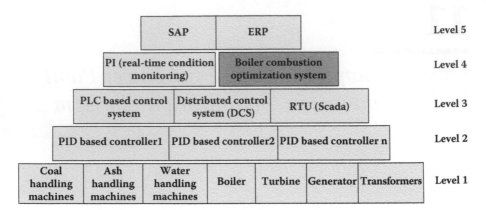

FIGURE 11.1
Conventional coal-fired power plant boiler combustion computer system.

system. Furthermore, the research integrates the latest technologies of high-performance computing, cloud computing, web services, and CORBA to provide a coal-fired power plant boiler combustion optimization platform based on the outputs of the research.

11.2 Building a Coal-Fired Power Plant Combustion Optimization System Supported by Online Learning Integrated with CFD in a Local Area Network

The software developed based on the methods of online learning integrated with CFD can be applied to improve coal-fired power plant boiler combustion efficiency locally. Figure 11.2 shows the solution applied locally in a coal-fired power plant. A general local area network in a coal-fired power plant is shown in Figure 11.2. The condition monitoring system and the boiler combustion optimization system are running on the network with a hardware isolator connected to the distributed control system and other intelligent device-based computer control networks.

The real-time data such as velocity of each burner and temperature of primary and secondary air are sent to the combustion optimization system, which is composed of a CFD-based server and online learning-based server. ANSYS Fluent 14.5 is applied in this research to simulate the combustion process and runs on the CFD-based server. The optimization and identification software is running on the online learning-based server. The distributed computing technology CORBA C++ is used to integrate both pieces of software as well as identify slagging distribution inside the boiler and optimize combustion process. Figure 11.2 shows that CORBA C++ is applied in the boiler combustion optimization system.

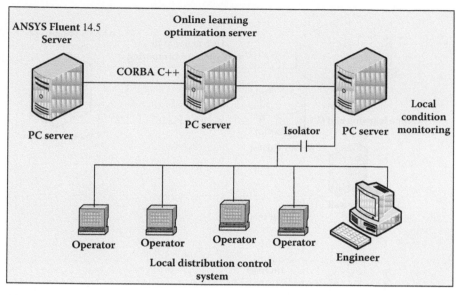

FIGURE 11.2
A coal-fired power plant boiler combustion optimization system supported by the technology
of online learning integrated with computational fluid dynamics is built in a local area network.

Because the CFD-based combustion simulation model is computing
intensive, the time for a coal-fired boiler combustion optimization is about
2–5 hours in our research in which the CFD based on combustion simulation and online learning-based optimization are running on a personal computer with 2.2 GHz central processing unit (CPU) and 8 GB random access
memory (RAM).

11.3 Using High-Performance Computer Technology to Build a Coal-Fired Power Plant Combustion Optimization System Supported by Online Learning Integrated with CFD

High-performance computer technology is applied to simulate highly
complex fluid dynamics in many areas of industry [142–144]. Figure 11.3
shows high-performance computing technology applied to simulate the
boiler combustion process; this can massively accelerate CFD-based boiler
combustion [142]. Figure 11.3 shows the CFD-based software ANSYS Fluent
running in a high-performance computer located in a university and online
learning-based optimization running on the server cluster. As shown in
Figure 11.2, the two servers are combined by CORBA C++. The technology
of web services can support the system to provide boiler combustion

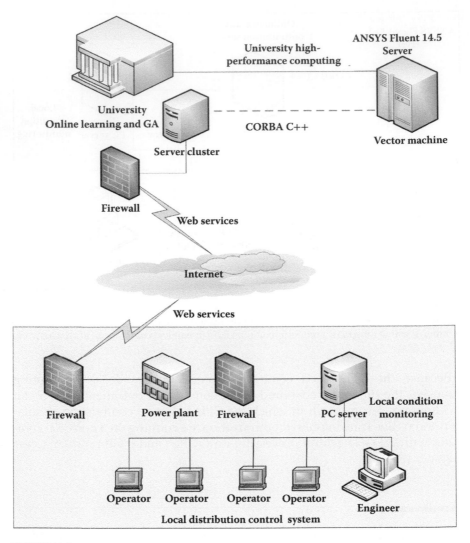

FIGURE 11.3
Coal-fired power plant boiler combustion optimization system based on integrating online learning with fluid dynamics supported by web services and high-perfomance computing.

optimization service to coal-fired power plants which can be connected to the system [145–148].

Compared with the coal-fired power plant boiler local combustion optimization system shown in Figure 11.2, the boiler combustion optimization system supported by high-performance computing can provide slagging identification and combustion optimization services to more than one power plant simultaneously.

11.4 Using Cloud-Computing Technology to Build a Coal-Fired Power Plant Combustion Optimization System Supported by Online Learning Integrated with CFD

Although the coal-fired boiler combustion optimization system supported by high-performance computing can provide optimization services to more than one coal-fired power plant, with many power plants simultaneously accessing the services provided by the system, the performance of the system can drastically decrease. However, cloud-computing technology can solve the problem and allow many power plants to access the services of the system simultaneously without any network access congestion [145–147]. Figure 11.4 shows a cloud-computing-supported coal-fired power plant boiler combustion optimization. A cloud, here named "optimization cloud," is created to integrate the high-performance computing resources from different universities or research centers in which high-performance computing resources are available. The CFD-based ANSYS Fluent 14.5 can run on each of the high-performance computers to simulate the combustion process. The online learning-based server can be connected to the optimization cloud and CORBA C++ is used to combine the server with the optimization cloud. Compared with the boiler combustion optimization system supported by high-performance computing, the cloud-technology-supported combustion optimization system can provide slagging distribution identification and combustion optimization services to many coal-fired power plants simultaneously without data congestion.

11.5 Integrating Online Learning Technology with CFD to Build a Coal-Fired Power Plant Boiler Combustion Optimization Platform Supported by High-Performance, Cloud-Computing, CORBA, and Web Services Technologies

In view that high-performance computing technology and Internet-related technologies, such as cloud computing, web services, and CORBA, have been applied in many areas [135–141], this research provides a solution on how to integrate online learning and CFD to improve coal-fired boiler combustion. The solution is shown in Figure 11.5, in which the coal-fired combustion optimization and slagging distribution identification platform is built by integrating online learning and CFD and supported by cloud-computing, CORBA, and web services technology. The platform is composed of three modules including an optimization cloud, database cloud, and combustion optimization. The three parts are linked by the Internet, and CORBA C++ is applied between the optimization cloud and combustion optimization module. The

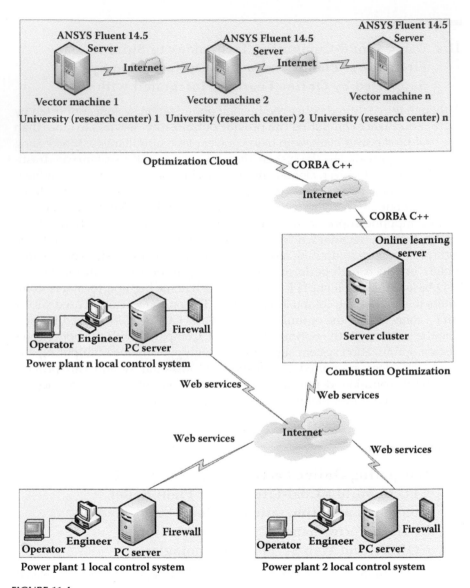

FIGURE 11.4
Cloud of high-performance computing resources applied to build a coal-fired power plant boiler combustion optimization system based on integrating online learning with fluid dynamics.

coal-fired power plants which require the services of slagging distribution identification and combustion optimization can be connected to the platform using web services technology with the Internet. The computing intensive combustion simulation is carried out in the optimization cloud module. The specific data of the coal-fired power plants which provide services are stored in the database cloud module.

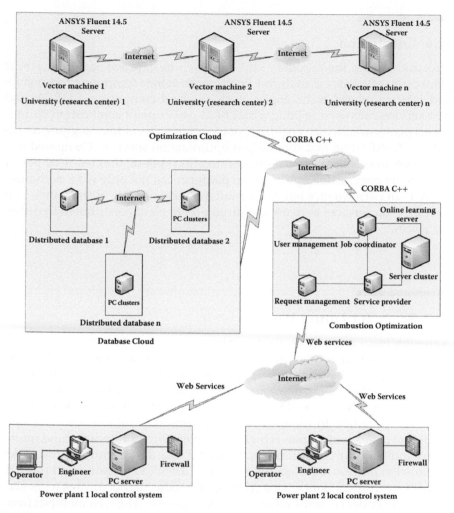

FIGURE 11.5
Internet-supported coal-fired power plant boiler combustion optimization platform.

As shown in Figure 11.5, the requirements for a coal-fired power plant are sent to the combustion optimization module and each request is processed in the module. For example, the geometry data, initial design data, and other specific data are collected into the database cloud module. The input parameters to be tuned are also decided by the system. Then the platform starts to work and results can return to the power plant in a real-time fashion. The web services technology is applied to support the access between the power plant local computer system and the combustion optimization module of the platform. For example, the optimal values for the input parameters can be continuously sent to the power plant local computer system to optimize the boiler combustion process.

11.6 Conclusion

Effectively using the outputs of this research to improve coal-fired power plant boiler efficiency is analyzed. High-performance computing technology and Internet-related technologies, such as cloud computing, CORBA, and web services, are used to build a coal-fired power plant combustion optimization platform in which coal-fired power plants can obtain slagging distribution identification and combustion optimization services. Compared to a local coal-fired boiler combustion optimization system, the coal-fired power plant boiler combustion optimization platform can not only work with massively high performance but also simultaneously provide optimization and identification services to many coal-fired power plants distributed in different places.

11.7 Scope for Future Works

Although fossil fuel power plants contribute a higher percentage of carbon emission than that of other industries, they still play a dominant role in the electricity generation industry. There is a great need to improve fossil fuel power plant efficiency in many countries of the world. Artificial intelligence (AI) technology-based fossil fuel combustion optimization methods are conventionally applied to improve fossil fuel power plant efficiency because fossil fuel boiler combustion is highly complex with nonlinearity and multi-input, multioutput characteristics. However, the data regarding the boiler combustion process and slagging are very difficult to measure and the lack of these instrument data means AI-based methods do not always work successfully in boiler combustion optimization because AI-based methods need measured data for more approximate model training.

This research has developed a novel method of integrating online learning, genetic algorithm (GA), and multiobjective optimization with CFD to improve fossil fuel power plant efficiency. Furthermore, the method can solve the slagging-related boiler combustion optimization problems on which the conventional AI-based fossil fuel power plant boiler combustion optimization methods cannot always work successfully. First, the method of integrating online learning with CFD can control the temperature field of flue gas and achieve much better control performance compared with proportional–integral–derivative (PID)-based control, which is widely used in power plants. Appropriately controlling the fields of flue-gas properties can significantly improve the fossil fuel power plant boiler combustion efficiency. Moreover, effectively controlling flue-gas temperature can avoid slagging caused by overheating.

Second, we discussed the method of integrating neural network–based multiobjective identification with CFD to identify the slagging distribution on the water walls of a fossil fuel tangential power plant boiler and encouraging results have been achieved. The results can not only be fed back to the control system for intelligent soot blowing but also conduct operation and predictive maintenance for a fossil fuel power plant boiler. Identifying slagging distribution inside a boiler is significant to improving the overall efficiency for a fossil fuel power plant. The software based on the research proposed method has been developed using CORBA C++, ANSYS Fluent 14.5, and Matlab to identify slagging distribution inside a coal-fired power plant boiler furnace.

Third, we looked at a method of integrating multiobjective optimization and CFD to optimize the boiler combustion process by tuning input parameters such as primary and secondary air temperature and velocity. The optimum results of the two objectives including heat transfer rate of furnace sides and temperature in the areas close to the sides of the furnace have been achieved. Compared with the corresponding values obtained without optimization, the proposed and developed method can keep the temperature of the areas close to the sides of furnace of the boiler within the ash melting temperature limit. However, the corresponding temperature value of the unoptimized flue gas is high enough for ash to melt and the ash melting will massively increase the slagging trends on the sides of the furnace of the boiler. Moreover, the heat transfer rate of the furnace sides in the optimized boiler combustion is similar or even a bit more than that of the unoptimized boiler combustion, although the temperature in the area close to the furnace sides is limited in the expected lower range. The software based on the proposed method has been developed using CORBA C++, ANSYS Fluent 14.5, and Matlab to optimize the fossil fuel boiler combustion process.

Finally, the research provides solutions on how to apply the research achievements in fossil fuel power plants and the general fossil fuel power plant local computer system is integrated in the solutions. The analysis for the solutions show that the boiler combustion optimization platform supported by high-performance computing and Internet-supported technology can provide more reliable and faster combustion process optimization to many power plants distributed in different places simultaneously. Furthermore, creating a higher performance computing cloud using cloud-computing technology to process CFD-based combustion simulation can make the fossil fuel power plant boiler combustion optimization platform more effective than the local boiler combustion optimization system based on the proposed method of integrating multioptimization with CFD.

The research proposes and has realized a novel method of integrating online learning with CFD to improve coal-fired power plant efficiency. In addition, the solutions on how to apply the developed method in a coal-fired power plant to improve boiler combustion efficiency are provided. Future research work to improve the solutions and extend the current research is discussed below.

1. *Research work on Internet-supported optimization*: With the advancement of Internet technology including cloud computing and web services, the Internet-supported computer system has become more reliable and faster. In addition, many successful industry practices show that cloud-computing technology can make an Internet-supported computer network more powerful by gathering distributed high-performance computing through the cloud. Compared with a local computer control system, Internet-supported computer control, identification, and optimization can be more robust, so there is a great need to do further research in this area.

2. *Extending the research-developed method to restrict the degradation of critical boiler equipment*: The work in this research finds that the reliability of the equipment inside the boiler can be badly influenced by inappropriate boiler combustion processes such as erosion caused by flue gas intensively and abnormally touching, corrosion caused by chemical reaction product such as NO_x and SO_y, and damage caused by overheating. Therefore, the research can be extended to not only optimize boiler combustion efficiency but also improve the reliability of equipment inside boiler such as the tubes of the water wall, superheaters, reheaters, and the economizer.

3. *Extending the research-developed method to improve condenser efficiency*: The appropriate condenser conditions can significantly improve the overall efficiency of a power plant. The work in this research can be extended to create a CFD-based model of a condenser process with steam inside of the condenser and a circle of water outside of the condenser. The model can not only monitor the condenser but also improve the condenser efficiency by optimizing the cooling water system including the cooling water pumps of the cooling tower system.

4. *Research on gas-fired power plant efficiency improvement*: Although this research focuses on a coal-fired power plant, it can also be extended to gas power plants to improve gas boiler efficiency and decrease carbon emissions.

References

1. W. Graus and E. Worrell, Comparison of efficiency fossil power generation, Central Research Institute of Electric Power Industry (CRIEPI), ECOFYS, Japan, p. 17, 2006 [Online]. Available at: http://www.ecofys.com/files/files/ecofyscomparison_fossil_power_efficiencyaug2006_02.pdf.
2. K. Sakcai, S. Morita, T. Yamamoto, and T. Tsumura, Design and operating experience on of the latest 1000-MW coal fired boiler, *Hitachi Review*, vol. 47, no. 5, pp. 183–187, 1998.
3. R. W. Bryers, Status and future of estimating slagging and fouling in steam generators, presented at *Practical Workshop on Impact of Coal Quality on Power Plant Performance*, Brisbane, Australia, Paper 11, pp. 1–38, 1994.
4. N. Spring, Putting combustion optimization to work, *Power Engineering*, vol. 3, no. 1, pp. 44–46, 2009.
5. U.S. Department of Energy, Power plant optimization demonstration projects, Clean Coal Technology, Department of Energy, The United States of America (September 2007), pp. 6–7 [Online]. Available at: http://www.netl.doe.gov/technologies/coalpower/cctc/topicalreports/pdfs/topical25.pdf.
6. G. P. Liu, *Nonlinear Identification and Control—A Neural Network Approach*, London, New York: Springer, pp. 178–192, 2001.
7. Y. Han, L. Xiu, Z. Wang, Q. Chen, and S. Tan, Artificial neural networks controlled fast valving in a power generation plant, *IEEE Transactions on Neural Networks*, vol. 8, no. 2, pp. 373–389, 1997.
8. W. Yang, K. Y. Lee, S. T. Junker, and H. G. Ayagh, Fuzzy fault diagnosis and accommodation system for hybrid fuel-cell/gas-turbine power plant, *IEEE Transactions on Energy Conversion*, vol. 25, no. 4, pp. 1187–1194, 2010.
9. V. Petridis, E. Paterakis, and A. Kehagias, A hybrid neural-genetic multi model parameter estimation algorithm, *IEEE Transactions on Neural Networks*, vol. 9, no. 5, pp. 862–876, 1998.
10. R. C. Morgans, C. J. Doolan, and D. W. Stephens, Derivative free global optimization of CFD simulations, *16th Australia Fluid Mechanics Conference*, Crown Plaza, Gold Coast, Australia, pp. 1432–1435, 2007.
11. M. Xu, X. He, J. L. T. Azevedo, and M. G. Crvalho, An advanced model to assess fouling and slagging in coal fired boilers, *International Journal of Energy Research*, vol. 26, no. 1, pp. 1221–1236, 2002.
12. T. Y. T. Lee and M. Mahalingam, Application of a CFD tool for system-level thermal simulation, *IEEE Transactions on Components, Package, and Manufacture Technology-Part A*, vol. 17, no. 4, pp. 564–572, 1994.
13. C. Ortiz, A. W. Skorek, M. Lavoie, and P. Bénard, Parallel CFD analysis of conjugate heat transfer in a dry-type transformer, *IEEE Transactions on Industry Applications*, vol. 45, no. 4, pp. 1530–1534, 2009.
14. B. D. J. Maynes, R. J. Kee, C. E. Tindall and R. G. Kenny, Simulation of airflow and heat transfer in small alternators using CFD, *IEEE Proceedings-Electric Power Applications*, vol. 150, no. 2, pp. 146–152, 2003.

15. J. Parry, R. B. Bornoff, P. Stehouwer, L. T. Driessen, and E. Stinstra, Simulation-based design optimization methodologies applied to CFD, *IEEE Transactions on Components and Packaging Technologies*, vol. 27, no. 2, pp. 391–397, 2004.

16. M. Fewster, Plant investigation of slagging at Callide Power Station, presented at *Practical Workshop on Impact of Coal Quality on Power Plant Performance*, Brisbane, Australia, paper 12, pp. 1–9, 1994.

17. M. Vinatoru, C. Maican, and G. Canureci, Heat exchange model for a power station boiler, presented at *IEEE International Conference Automation Quality and Testing Robotics*, Cluj-Napoca, Romanian, pp. 26–31. 2012.

18. E. An and H. Zhou, Digital simulation for coal fired power plant combustion, *Journal of Tongji University (Natural Science)*, vol. 32, no. 8, pp. 1042–1045, 2004.

19. H. K. Versteeg, *An Introduction to Computational Fluid Dynamics-the Finite Volume Method*, 2nd edition, London: Pearson Education, pp. 343–416, 2007.

20. C. E. Baukal, J. V. Gershtein, and X. Li, *Computational Fluid Dynamics in Industrial Combustion*, New York: CRC Press, pp. 139–143, 2001.

21. ANSYS Inc., *ANSYS Fluent Theory Guide*, United States: ANSYS, Inc., pp. 121–136, 2012.

22. Y. A. Cengel and M. A. Boles, *Thermal Dynamics—An Engineering Approach*, 7th edition in SI units, London: High Education, pp. 689–833, 2011.

23. J. Tu, G. H. Yeoh, and C. Liu, *Computational Fluid Dynamics—A Practical Approach*, New York: Elsevier, pp. 180–221, 2008.

24. Sector Policies and Programs Division, Greenhouse gas emissions from coal-fired electric generating units, Office of air quality planning and standards, U.S. Environmental Protection Agency, North Carolina (October 2010).

25. IEA Statistics, *CO2 Emissions from Fuel Combustion*, 2014 edition, Paris, France: International Energy Agency.

26. CEA, Installed capacity, Ministry of Power, Govt. of India, New Delhi (April 2014) [Online]. Available at: http//:www.cea.nic.in.

27. CEA, Growth of electricity sector in India from 1947–2013, Ministry of Power, Govt. of India, New Delhi (July 2013) [Online]. Available at: http//:www.cea.nic.in.

28. EPRI, Program on technology innovation: Electricity use in the electric sector, opportunities to enhance electric energy efficiency in the production and delivery of electricity, Electric Power Research Institute, Technical report, pp. 2-1 to 2-4, 2011.

29. VGB Powertech, Compendium best practices for coal-based power plants in Germany, pp. 106, 139 (August 2014).

30. CERC, Notification for terms and conditions of tariff regulations 2009–2014, vide Order No.L-7/145(160)/2008-CERC dt, pp. 48–49 (19th January 2009).

31. R. P. Mandi and U. R. Yaragatti, Control of CO_2 emission through enhancing energy efficiency of auxiliary power equipment in thermal power plant, *International Journal of Electrical Power & Energy*, vol. 62, pp. 774–752, 2014.

32. R. P. Mandi, S. Seetharamu, and U. R. Yaragatti, Energy efficiency improvement of auxiliary power equipment in thermal power plant through operational optimization, presented at *Proceedings of IEEE International Conference on Power Electronics, Drives and Energy Systems (PEDES)*, published in IEEE Explorer, Bengaluru, India, co-organized by CPRI, Bangalore, at National Science Seminar Complex, CSIC, IISc, Bangalore, December 16–19, 2012.

33. R. P. Mandi, S. Seetharamu, and U. R. Yaragatti, Enhancing energy efficiency of boiler feed pumps in thermal power plants through operational optimization and energy conservation, *International Research Journal of Power and Energy*, vol. 1, no. 2, pp. 013–016, 2014.

34. R. Sethu Rao and G. A. Purushotam, Fuel & efficiency—coal combustion improvement, coal beneficiation and coal blending, presented at *Proceedings of National Symposium on Energy Conservation Measures in Generating Sector*, Bangalore, November 17–18, pp. I-54–I-64, 2005.

35. V. Saravanan, R. K. Kumar, R. Suprabha, and S. Seetharamu, A CFD study on the blended coal combustion in a typical 210 MW Indian boiler, presented at *Proceedings of National Conference on 'Technological Advances for New Power Generating Units and for Performance Enhancement of Present Plants'*, Bangalore, November 18–19, pp. I-83–I-91, 2010.

36. S. K. Storm, J. Guffre, and A. Zucchelli, Advancements with regenerative air heater design, performance and reliability, POWER-GEN Europe, ID-1546, June 7–9, FIERA, MILANO–MILAN–ITALY, pp. 1–16, 2011.

37. G. R. Venkataraman and K. Mariraj Anand, Energy conservation improvements in air pre-heaters of boilers, presented at *Proceedings of National Symposium on Energy Conservation Measures in Generating Sector*, Bangalore, November 17–18, pp. V-1–V-5, 2005.

38. A. P. Chikkatur and A. D. Sagar, Cleaner power in India: Towards a clean-coal-technology roadmap, Energy technology innovation policy, John F. Kennedy School of Government, Harward University, Discussion paper 2006-07, 2007.

39. Vir J. Koul, Metallic rotary gravimetric raw coal feeders for power plant PC boilers, presented at *Proceedings of National Conference on 'Technological Advances for New Power Generating Units and for Performance Enhancement of Present Plants'*, Bangalore, November 18–19, pp. I-28–I-36, 2010.

40. S. G. Chandwani, F. Turoni, R. Kock, M. Haug, and M. Schreiber, Online coal flow measuring and balancing (control) and online coal fineness measuring system for combustion optimization in a thermal boiler, presented at *Proceedings of National Conference on 'Technological Advances for New Power Generating Units and for Performance Enhancement of Present Plants'*, Bangalore, November 18–19, pp. I-39–I-41, 2010.

41. R. P. Mandi, Enhancing energy efficiency of out-lying auxiliary equipment through energy conservation in TPS, presented at *Proceedings of National Workshop on Energy Conservation for Power Engineers*, at PMC, Noida, New Delhi, Organized by NTPC, October 20–21, pp. 16–24, 2005.

42. M. M. El-Wakil, *Powerplant Technology-International Edition 1984*, McGraw-Hill Book Company, Singapore, pp. 260–306, 1984.

43. L. Backer, W. M. Wurtz, Why every air cooled steam condenser needs a cooling tower, Paper No:TP03-01, presented at *Annual Conference of CTI, 2003, Cooling Technology Institute Annual Conference*, San Antonio, Texas, February 10–13, 2003.

44. Cooling tower thermal design manual, http://myhome.hanfos.com.

45. Modulus counterflow cooling tower, http://www.marleyct.com.

46. Replacing an oversized and under-loaded electric motor, http://www.oit.doe.gov/bestpractices/motors.

47. H. W. Penrose, *A Novel Approach to Electric Motor System Maintenance and Management for Improved Industrial and Commercial Uptime & Energy Costs*, 2nd edition, Old Saybrook CT: H. W. Penrose. 2001.

48. R. P. Mandi, R. K. Hegde, and S. N. Sinha, Performance enhancement of cooling towers in thermal power plants through energy conservation, published at *International Conference on POWERTECH 2005*, St. Petersburg, Russia, Organized by IEEE Power Engineering Society, June 27–30, Paper No. 461, 2005.

49. A. J. Wood, B. F. Wollenberg, *Power Generation Operation and Control*, New York: Wiley, 1984.

50. S. Moaveni, *Finite Element Analysis Theory and Application with ANSYS*, 3rd edition, London, Person Education, pp. 405–745, 2008.

51. A. W. Ordys, A. W. Pike, M. A. Johnson, R. M. Katebi, and M. J. Grimle, *Modeling and Simulation of Power Generation Plants*, London: Springer–Verlag, pp. 118–212, 1994.

52. J. Li., J. Liu and Y. Niu, On-line self-optimizing control of coal fired boiler combustion system, presented at *Proceedings of the Third International Conference on Machine Learning and Cybernetics*, Shanghai, China, pp. 564–567, 2004.

53. R. Balakrishnan, Advanced Topic in Engineering I, report, The University of Queensland, pp. 21, 2010.

54. R. Lohner, *Applied CFD Techniques—An Introduction Based on Finite Element Methods*, Chichester: John Wiley & Sons Ltd, pp. 1–3, 2001.

55. A. W. Ordys, *Modelling and Simulation of Power Generation Plants*, Springer, London, pp. 117–200, 1994.

56. C. Liu, J. Lin, Y. Nui and W. Liang, Nonlinear boiler model of 300 MW power unit for system dynamic performance studies, presented at *Proceedings ISIE, IEEE International Symposium on Industrial Electronics*, vol. 2, June 12–16, pp. 1296–1300, 2001.

57. D. Flynn, *Thermal Power Plant Simulation Control*, The Institute of Electrical Engineers, London, pp. 161–177, 2003.

58. Q. Chen, X. Yan, J. Jiang, and G. Sun, Application of gray system in boiler slagging properties in power plant, presented at *International Conference Energy and Environment Technology*, Guilin, China, 2009.

59. E. B. Ziv, Y. Berman, R. S. Perelman, M. Korytnyi, E. B. Davidson, and B. Chudnovsky, Fouling formation in 575 MW tangential-fired pulverized-coal boiler, *Journal of Engineering for Gas Turbines and Power*, vol. 132, no. 12, pp. 1–7, 2010.

60. C. Bao, N. Cai, and E. Croiset, A multi-level simulation platform of natural gas internal reforming solid oxide fuel cell-gas turbine hybrid generation system—Part II. Balancing units model library and system simulation, *Journal of Power Sources*, vol. 196, no. 20, pp. 8424–8434, 2011.

61. N. Tsioumanis, J. G. Brammer, J. Hubert, Flow process in a radiant tube burner: Combusting flow, *Energy Conversion and Management*, vol. 52, no. 1, pp. 2667–2675, 2011.

62. C. Ghenai and I. Janajreh, CFD analysis of the effects of co-firing biomass with coal, *Energy Conversion and Management*, vol. 51, no. 8, pp. 1694–1701, 2010.

63. N. Punbusayakul, J. Charoensuk, and B. Fungtammasan, Modified sulfation model for simulation of pulverized coal combustion, *Energy Conversion and Management*, vol. 47, no. 3, pp. 253–272, 2006.

64. T. Y. T. Lee, B. Chambers, and M. Mahalingam, Application of CFD technology to electronic thermal management, *IEEE Transactions on Components, Packaging, and Manufacturing Technology, Part B: Advanced Packaging*, vol. 18, no. 3, pp. 511–520, 2002.

65. G. D'Rrrico, Prediction of the combustion process and emission formation of a bi-fuel s.i. engine, *Energy Conversion and Management*, vol. 49, no. 11, pp. 3116–3128, 2008.

66. S. Du, W. Chen, and J. Lucas, Performances of pulverized coal injection in blowpipe and tuyere at various operational conditions, *Energy Conversion and Management*, vol. 48, no. 7, pp. 2069–2076, 2007.

67. Y. Yu, D. Sun, K. Wu, Y. Xu , H. Chen, and X. Zhang, CFD study on mean flow engine for wind power exploitation, *Energy Conversion and Management*, vol. 52, no. 6, pp. 2355–2359, 2011.

68. G. Zhou, L. Chen L, and J. Seaba, CFD prediction of shunt currents present in alkaline fuel cells, *Journal of Power Sources*, vol. 196, no. 20, pp. 8180–8187, 2011.

69. T. Y. T. Lee and M. Mahalingam, Application of a CFD tool for system-level thermal simulation, *IEEE Transactions on Components, Packaging, and Manufacturing Technology, Part A*, vol. 17, no. 4, pp. 564–572, 2002.

70. M. G. Camprubi, H. Jasak, and N. Fueyo, CFD analysis of cooling effects in H2-fed solid oxide fuel cells, *Journal of Power Sources*, vol. 196, no. 17, pp. 7290–7301, 2011.

71. A. W. Date, *Introduction to Computational Fluid Dynamics*, New York: Cambridge University Press, pp. 273–283, 2005.

72. E. M. Alawadhi, *Finite Element Simulation Using ANSYS*. London: CRC Press Taylor & Francis Group, pp. 183–188, 2010.

73. K. Y. Lee, J. H. V. Sickel, J. A. Hoffman, and W. Jung, Controller design for a large-scale ultra-supercritical once-through boiler power plant, *IEEE Transactions on Energy Conversion*, vol. 25, no. 4, pp. 1063–1070, 2010.

74. G. Catlin, S. G. Advani and A. K. Prasad, Optimization of polymer electrolyte membrane fuel cell flow channels using a genetic algorithm, *Journal of Power Sources*, vol. 196, no. 22, pp. 9407–9418, 2011.

75. S. Bashash, S. J. Moura, J. C. Forman, and H. K. Fathy, Plug-in hybrid electric vehicle charge pattern optimization for energy cost and battery longevity, *Journal of Power Sources*, vol. 196, pp. 541–549, 2011.

76. M. Pouraghaie, K. Atashkari, S. M. Besarati, and N. Narimanzadeh, Thermodynamic performance optimization of a combined power/cooling cycle, *Energy Conversion and Management*, vol. 51, no. 1, pp. 204–211, 2010.

77. Y. Li, J. Shen, and J. Lu, Constrained model predictive control of a solid oxide fuel cell based on generic optimization, *Journal of Power Sources*, vol. 196, no. 14, pp. 5873–5880, 2011.

78. E. Assareh, M. A. Behrang, M. R. Assari, and A. Ghanbarzadeh, Application of PSO (particle swarm optimization) and GA (genetic algorithm) techniques on demand estimation of oil in Iran, *Energy*, vol. 25, no. 12, pp. 5223–5229, 2010.

79. T. Opher and A. Ostfeld, A coupled model tree (MT) genetic algorithm (GA) scheme for biofouling assessment in pipelines, *Water Research*, vol. 45, no. 18, pp. 6277–6288, 2011.

80. A. Egea, R. Marti, and J. R. Banga, An evolutionary method for complex-process optimization, *Computers and Operations Research*, vol. 37, no. 2, pp. 316–319, 2010.

81. S. G. Dukelow, *The Control of Boilers,* 2nd edition, The United States of America: The Instrumentation, Systems, and Automation Society, pp. 253–256, 1991.

82. H. Zhou, L. Zheng, K. Cen, Computational intelligence approach for NOx emissions minimization in a coal-fired utility boiler, *Energy Conversion and Management,* vol. 51, no. 3, pp. 580–586, 2009.

83. B. Zarenezhad and A. Aminian, Accurate prediction of the dew points of acidic combustion gases by using an artificial neural network model, *Energy Conversion and Management,* vol. 52, no. 2, pp. 911–916, 2011.

84. H. Rusinowski and W. Stanek, Neural modeling of steam boilers, *Energy Conversion and Management,* vol. 48, no.11, pp. 2802–2809, 2007.

85. J. S. Chandok, I.N. Kar, and S. Tuli, Estimation of furnace exit gas temperature (FEGT) using optimized radial basis and back-propagation neural networks, *Energy Conversion and Management,* vol. 49, no. 8, pp. 1989–1998, 2008.

86. J. Zhang, F. Zhao, and C. Dong, CFD studies on the air flow in a double-grate biomass fired boiler, presented at *International Conference Digital Manufacturing and Automation (ICDMA),* Changsha, China, 2010.

87. F. Xiao and J. D. Mcalley, Power system risk assessment and control in a multi objective framework, *IEEE Transactions on Power Systems,* vol. 24, no. 1, pp. 78–85, 2009.

88. P. Murugan, S. Kannan, and S. Baskar, Application of NSGA-II algorithm to single-objective transmission constrained generation expansion, *IEEE Transactions on Power Systems,* vol. 24, no. 4, pp. 1790–1796, 2009.

89. J. H. Kim, J. H. Han, Y. H. Kim, S. H. Choi, and E. S. Kim, Preference-based solution selection algorithm for evolutionary multi objective optimization, *IEEE Transactions on Evolutionary Computation,* vol. 16, no. 1, pp. 20–33, 2012.

90. X. Meng and J. Ni, Research of active vibration control optimal disposition based on MIGA and NSGA-II, presented at *Sixth International Conference Natural Computation,* Yantai, China, 2010.

91. V. Baghel, G. Panda, P. Srihari, K. Rajarajeswari, and B. Majhi, An efficient multi-objective pulse radar compression technique using RBF and NSGA-II, presented at *Conference on Nature & Biologically Inspired Computing,* Bhubaneswar, India, 2009.

92. J. Zhao, J. Fu, H. Chen, and L. Zhang, Decentralized PID controllers of steam-turbine generator set based on probabilistic robust method, presented at *International Conference on Future Information Technology and Management Engineering,* Changzhou, China, 2010.

93. M. B. Fallanhpour, K. D. Hemmati, A. Poumohammand, and A. Golmakani, Multi objective optimization of a LNA using genetic algorithm based on NSGA-II, presented at *International Conference Electrical Engineering and Informatics,* Bandung, Indonesia, 2011.

94. C. Dong, Y. Yang, and R. Yang, Numerical modeling of the gasification based biomass co-firing in a 600 MW pulverized coal boiler, *Applied Energy,* vol. 87, no. 9, pp. 2834–2838, 2009.

95. Y. Xu, Q. Jin, and J. Yuang, CFD simulations of gate leaves design in the SCR-DeNOx facility for coal-fired power plant, presented at *30th Chinese Control Conference,* Yantai, China, 2011.

96. J. Zhou, W. Han, and X. Liu, Design and experiment of silencer for discharging waste water of high temperature and pressure in nuclear power plant, presented at *Asia-Pacific Power and Energy Engineering Conference,* Chengdu, China, 2010.

97. D. Shen and P. Zhao, Computational fluid dynamics-based simulation of optimal design in power plant flue-gas denitrification, presented at *International Conference Communications and Control,* Zhejiang, China, 2011.

98. Y. Fan, P. Si, and B. Bai, Analysis and numerical simulation on an improved oxygen-enriched burner, presented at *Asia-Pacific Power and Energy Engineering Conference,* Chengdu, China, 2010.

99. X. Dang and H. Hu, Study of numerical simulation and experimental of gas flow distribution of electric composite filter bag in 300MW power unit in coal-fired power plant, presented at *4th International Conference Bioinformatics and Biomedical Engineering,* 2010, Chengdu, China, 2010.

100. X. Wang, M. Wang, and Y. Wei, Simulation study on data of coal fired power plant boiler experiment with cold state, *Energy Research and Utilization,* vol. 4, no. 1, pp. 5–8, 2007.

101. L. Juniper, *Thermal Coal Technology,* Australia, Department of Mines and Energy, Queensland, Government, pp. 71–128, 2000.

102. P. Chattopadhyay, *Boiler Operation Engineering Questions and Answers,* 2nd edition, New York, McGraw Hill, pp. 941–998, 2001.

103. Dual loop control with disturbance determination, taking the example of steam temp. Control and simulation of the transition function, SIEMENS.

104. C. N. Ning and C. N. Liu, Effects of temperature control on combined cycle unit output response, *IEEE Region 10 Conference,* November 14–17, pp. 1–4, 2006.

105. Control of Steam Pressure from Steam Generators in Power Plant Units, SIEMENS, Aust Electric Stanwell Power Station, pp. 1–22.

106. Feedwater supply for a once through steam generator with a circulating unit in steam power plants, SIEMENS.

107. G. Prasad, E. Swidenbank, and B. W. Hogg, A neural net model-based multivariable long-range predictive control strategy applied in thermal power plant control, *IEEE Transactions on Energy Conversion,* vol. 13, no. 2, pp. 176–182, 1998.

108. Y. Y. Nazaruddin, A. N. Azizi, and W. Sudibjo, Improving the performance of industrial boiler using artificial neural network modeling and advanced combustion control, presented at *International Conference on Control and Automation and System,* Seoul, Korea, pp. 1921–1926, 2008.

109. E. Bar-Ziv, Y. Berman, R. Saveliev, M. Perelman, E. Korytnyi, B. Davidson, and B. Chudnovsky, Fouling formation in 575 MW tangential-fired pulverized-coal boiler, *Journal of Engineering for Gas Turbines and Power,* vol. 132, no. 1, pp. 1–7, 2010.

110. Z. Ma, F. Iam, P. Lu, R. Sears, L. Kong, A. S. Rokanuzzaman, D. P. McCollor, and S. A. Benson, A comprehensive slagging and fouling prediction tool for coal-fired boilers and its validation/application, *Fuel Processing Technology,* vol. 88, no. 11, pp. 1035–1043, 2007.

111. M. Manickam, M. P. Schwarz, and M. J. Mcintosh, CFD analysis of erosion of bifurcation duct walls, presented at *Second International Conference on CFD in the Mineral and Process Industry,* Melbourne, Australia, pp. 243–248, 1999.

112. Z. Tian, P. J. Witt M. P. Schwarz, and W. Yang, Numerical modeling of brown coal combustion in a tangentially fired furnace, presented at *Seventh International Conference on CFD in the Mineral and Process Industry*, Melbourne, Australia, December, pp. 1–8, 2009.

113. D. Kerlick, E. Dillon, and D. Levine, Performance testing of a parallel multi-block CFD Solver, *International Journal of High Performance Computing Applications*, vol. 15, no. 1, pp. 22–35, 2001.

114. D. M. Jaeggi, G. T. Parks, T. Kipouros, and P. J. Clarkson, The development of a multi-object Tabu Search algorithm for continuous optimization problems, *European Journal of Operational Research*, vol. 185, no. 3, pp. 1192–1212, 2008.

115. W. Shen, Q. Hao, H. J. Yoon, and D. H. Norrie, Applications of agent-based system in intelligent manufacturing: An updated review, *Advanced Engineering Informatics*, vol. 20, no. 4, pp. 415–431, 2006.

116. Y. Zhu and Z. Gao, Experimental research on fouling characteristics of heating surfaces in boiler burning Shenhua coal, presented at *International Conference on Power and Energy Engineering Conference*, Wuhai, China, 2009.

117. K. M. Pssino, *Biomimicry for Optimization, Control, and Automation*, London: Springer, pp. 549–573, 2005.

118. I. T. Nabney, *Netlab Algorithms for Pattern Recognition*, New York: Springer, pp. 191–223, 2002.

119. Y. Zhang, X. Pu, and Q. Yang, The intelligent soot-blowing programmable control system which is based on improved Elman neural network, presented at *3rd International Conference Measuring Technology and Mechatronics Automation*, Shanghai, China, 2011.

120. A. Ghaffari, S. A. A. Moosavian, and A. Chaibakhsh, Experimental fuzzy modeling and control of a once-through boiler, presented at *IEEE International Conference Mechatronics and Automation*, Niagara Falls, Ontario, Canada, 2005.

121. X. Liu, X. Tu, G. Hou, and J. Wang, The dynamic neural network model of an ultra-super-critical steam boiler unit, presented at *American Control Conference*, San Francisco, CA, pp. 2474–2479, 2011.

122. S. Thompson and N. Li, Boiler fouling, monitoring and control, *Computing & Control Engineering Journal*, vol. 3, no. 6, pp. 282–286, 1992.

123. L. Ma, Y. Ma, and K. Y. Lee, An intelligent power plant fault diagnostics for varying degree of severity and loading conditions, *IEEE Transactions on Energy Conversion*, vol. 25, no. 2, pp. 546–554, 2010.

124. S. Tan and C. Lim, Application of an adaptive neural network with symbolic rule extraction to fault detection and diagnosis in a power generation plant, *IEEE Transactions on Energy Conversion*, vol. 25, no. 2, pp. 369–377, 2004.

125. E. Odvarka, N. L. Brown, M. Shanel, S. Narayanan, and C. Ondrusek, Thermal modelling of water-cooled Axial-Flux Permanent Magnet Machine, presented at *5th IET International Conference Power Electronics, Machines and Drives*, Brighton, UK, 2010.

126. B. Jayashankara, V. Ganesan, Effect of fuel injection timing and intake pressure on the performance of a DI diesel engine—A parametric study using CFD, *Energy Conversion and Management*, vol. 51, no. 10, pp. 1835–1848, 2010.

127. M. G. Camprubi, H. Jasak, and, N. Fueyo, CFD analysis of cooling effects in H2-fed solid oxide fuel cells, *Journal of Power Sources*, vol. 196, no. 17, pp. 7290–7301, 2011.

128. J. Saranagapani, *Neural Network Control of Nonlinear Discrete-Time System*, New York: Taylor & Francis, United States, pp. 276–337, 2006.

129. C. M. Bishop, *Neural Network for Pattern Recognition*, Oxford, Oxford University Press, pp. 145–147, 1995.

130. F. Li, D. Wei, and J. Ma, The digital simulation of decreasing NOx emission based on classified coal fired boiler combustion, *Journal of Fuel Chemistry and Technology*, vol. 32. no. 5, pp. 537–541, 2004

131. L. Zheng, H. Jia, and M. Yu, Prediction of NOx concentration from coal combustion using LS-SVR, presented at *Bioinformatics and Biomedical Engineering, 4th International Conference*, Chengdu, China, 2010.

132. S. Batmunkh, T. S. Tseyen-Oidov, and Z. Battogtokh, Survey to develop standards on air polluting emissions from the power plants, presented at *Third International Forum Strategic Technologies*, Novosibirsk-Tomsk, Russia, 2008.

133. K. K. Kuo, *Principles of Combustion*, 2nd edition, Trenton: John Wiley & Sons Ltd, US, pp. 215–297, 2005.

134. H. G. Beyer and H. Deb, On self-adaptive features in real parameter evolutionary algorithms, *IEEE Transactions on Evolutionary Computation*, vol. 5, no. 3, pp. 250–268, 2001.

135. Y. Joo, H. Jun, C. Jug, and J. Soon, CORBA based core middleware architecture supporting seamless interoperability between standard home network middleware, *IEEE Transaction on Consumer Electrics*, vol. 49, no. 3, pp. 581–586, 2003.

136. D. Levy and A. Liu, Evaluating overhead and predictability of a real-time CORBA system stack of Linux rather than in TAO itself, presented at *37th Annual International Conference System Sciences*, Hawaii, US, 2004.

137. C. Rebeiro and R. Pitchiah, SCADA with fault tolerant CORBA on fault tolerant LANE ATM, presented at *19th IEEE International Parallel and Distributed Processing Symposium*, Bangalore, India, 2005.

138. S. Gong and Y. Liu, Study and design of integrated transmission network management system based on CORBA and web, presented at *International Conference Industrial Control and Electronics Engineering*, Xi'an, China, 2012.

139. L. Guo, X. Wang, and P. Kou, Design of distributed network management system based on web and CORBA, presented at *International Conference Internet Technology and Applications*, Wuhan, China, 2010.

140. X. Lai and J. Le, Power system simulation platform based on CORBA technology, presented at *International Conference Electronic and Mechanical Engineering and Information Technology*, Harbin, China, 2011.

141. M. Ferber, T. Rauber, and S. Hunold, Combining object-oriented design and SOA with remote objects over web services, presented at *8th Web Services IEEE European Conference*, Ayia Napa, Cyprus, 2010.

142. F. Lu, J. Song, X. Cao, and X. Zhu, Acceleration for CFD applications on large GPU clusters: An NPB case study, presented at *6th International Conference Computer Sciences and Convergence Information Technology*, Seogwipo, South Korea, 2011.

143. D. S. Roman, G. Sutter, and S. L. Buedo, High-level languages and floating-point arithmetic for FPGA-based CFD simulation, *IEEE Design & Test of Computers*, vol. 28, no. 4, pp. 28–36, 2011.

144. X. Wang, G. Xing, and C. Lin, Leveraging thermal dynamics in sensor placement for overheating server component, presented at *International Green Computing Conference*, San Jose, USA, 2012.

145. K. M. Sim, Agent-based cloud computing, *IEEE Transaction Services Computing*, vol. 5, no. 4, pp. 564–5, 2012.
146. S. Malik, F. Huet, and D. Caromel, *Cooperative Cloud Computing in Research and Academic Environment Using Virtual Cloud*, Islamabad, Pakistan, 2012.
147. Y. Jadeja and K. Modi, Cloud computing—Concepts, architecture and challenges, presented at *International Conference Computing, Electronics and Electrical Technologies*, Kherva, India, 2012.
148. F. B. Shaikh and S. Haider, Security threats in cloud computing, presented at *International Conference Internet Technology and Secured Transactions*, Islamabad, Pakistan, 2011.

Index

Printed and bound by CPI Group (UK) Ltd, Croydon, CR0 4YY

24/10/2024

01778302-0009